International Praise for
Holding IT Together

In *Holding IT Together,* Robert MacNeil Christie forces us to recognise that as we approach environmental tipping points, we are equally hurtling towards the end of Industrial Civilisation. The Holocene, and the Goldilocks period, enjoyed by only a privileged segment of society, is effectively over. Political-economic and financial powers have succeeded in taking away the democratic power of the people they are meant to represent, to the detriment of the general populace. This leads to social chaos. The source of our hope is our agency—by mustering our political will and cultural creativity, leveraging true democratic politics, science and indigenous knowledge to create a pathway way to our human survival. I highly recommend this book, as an eye opener to the systemic levers that are causing our collective inaction, but also the power we have to turn this juggernaut around quickly.

~ Antoinette Vermilye, Co-Founder of She Changes Climate; Co-Founder of the Gallifrey Foundation, Geneva Area, Switzerland

Holding IT Together is somewhat rare in the environmental space today: an honest and accessible exploration of why we are on the precipice of ecological and social collapse, and what we need to do to retreat from the edge. What you won't find in these pages is unrealistic and unfounded techno-optimism, nor is there the hero-worshipping of today's so-called "leaders" (both business and political) to "save us" that we see so often in the environmental movement.

What you will find is a candid and well-researched analysis of social systems, the harms of growth-based economics, the role of imperialism in maintaining globalised corporate states, the flip-side of 'progress', the need for degrowth and ecological societies and the role of bona fide democracy in getting there.

Holding IT Together is an important and necessary read for anyone who is looking for a genuine understanding of our predicament and what we can do from here.

~ Erin Remblance, widely followed commentator, analyst, and activist on Climate Chaos, Economic Degrowth, and Social Ecology, Sydney, Australia

Robert Christie has delivered! On September 16, 2014, on his blog *TheHopefulRealist.com*, he posed a simple, yet critical question: Climate Science or Social Science? The chasm continues between the facts of climate change and the inability (or unwillingness) to transfer that reality into focused social action and change: Science had spoken and it was now the people's turn. *Holding IT Together*, asks us to be more aware of and to utilize our collective-intersectional-agency in order to survive. We don't need a makeover but a do-over—a complete reset in how we understand and can ultimately survive, in the ONE future we all will share. As the saying goes, all politics are local—and so is our survival—we must all do our part to dismantle the current normalization of human and environmental destruction and the societal structures that support it.

~ **Derrick Jones, PhD, Director and Community Liaison for the UCSC Oakes College Community-Based Action Research and Advocacy Program, University of California, Santa Cruz**

Holding IT Together is quite interesting and educative at the same time. Dr. Christie's decades of research and expertise shine through in this important work; his anecdotes and observations are intertwined throughout the chapters, making the content feel authentic and relatable. His writing style is conversational and accessible, making the content feel approachable and engaging.

~ **Olatomide Ojajune, author of *Green Earth Green Life*, Host on *Talk Climate Africa*, Forester and Environmental Advocate, Abuja, Federal Capital Territory, Nigeria**

Incisive critique of the extreme hierarchy of capital and power we live within, how things got this way and how maladaptive it remains for us and the planet.

~ **James F. Richardson, PhD, author of *Our Worst Strength: American Individualism and Its Hidden Discontents***

Holding IT Together is timely and compelling. Drawing on his distinguished career as a sociologist, Robert MacNeil Christie argues persuasively that we are faced with a convergence of environmental and social crises. These result from over-exploitation of natural resources and reliance on unsustainable economic models. He maintains that global society has now reached a tipping point: Earth can no longer sustain current levels of extraction, consumption and waste. In response, Dr. Christie proposes innovative ideas on how we can avert a global catastrophe. This thought-provoking book is a must read for policy makers and concerned citizens.

~ **Mark L. Asquino, U.S. Ambassador (Ret.)**
and author of *Spanish Connections: My Diplomatic*
Journey from Venezuela to Equatorial Guinea

Social psychologist Robert Christie brings to the climate discussion well-articulated points about behavioral and institutional changes necessary to build a viable ecological civilization, making a strong case that our existential predicament will not be resolved by technical fixes by global power elites such as the "Davos Men." The answers are not for sissies and their urgency calls for immediate action.

~ **Earl Kessler, author of** *Letters from Alfonso: Learning to Listen*;
Advisor: Shelter and Settlements Team Bureau for Humanitarian
Assistance 2019-2023; Deputy Executive Director Asian Disaster
Preparedness Center 2003-2006; Consultant: Asian Development
Bank, World Bank, UN-HABITAT; Lead Author: Climate Resilient
Cities Primer, World Bank, 2009; Director of the Regional Urban
Development Office/South Asia/USAID 1993-1998

This is an excellent and thought provoking read. As one who makes her living investing in and teaching about stocks and commodities, I note that *Holding IT Together* brings up several excellent points related to macroeconomics. For example, one kind of person Dr. Christie discusses believes technology and technological innovation will fix everything, and made huge amounts of money buying tech stocks. This is likely to continue. Tech boom pessimists amassed huge losses, making the nihilists only look "fringe". The stock market itself may reflect "hopeful realists", who can only hope that this book cements the notion that we collectively know what looks like the best way out of this mess.

~ **Michele 'Mish' Schneider, Chief Strategist** *MarketGauge.com,*
and financial commentator on many television and radio shows

Robert MacNeil Christie's new book, *Holding IT Together*, articulates the twist and turns of the past that culminate in our complex present. Dr. Christie doesn't hold back on the dire, urgent situation we find ourselves in and how difficult and important it is that we act from our best selves. We must collaborate in seeking a much smaller, healthier, and kinder world. Robert offers lots of principles, practices and tips to get us there. He reminds us there is no secret formula. I suggest reading H.I.T. and doing it. Let's communicate and rebuild together, as Christie suggests.

~ **Pam Pence, MBA, CPA, and Executive Director,**
Stable Planet Alliance

Robert MacNeil Christie blends a rich mix of personal history, exhaustive research and perceptive analysis to make a strong case for a sustainable natural society, and he offers a blueprint for building a more livable world. He indeed shows us the necessity for *Holding IT Together*.

~ **Jack Heinsius, Professor of Business**
and Political Science Emeritus, Modesto Junior College

As a social psychologist, Robert Christie makes a strong case with his book *Holding IT Together* for a serious discussion on the subject of climate change. His articulation on the significance of human behavior and the need for change is clear. This is crucial if our world is to continue to remain a viable medium for our existence. It is on us, and we need to heed Roberts words if we are to survive what our behavior has caused.

~ **Mark Van Patten, author of** *Moonshine and Watermelons*
and other Ozark Tales, **retired fisheries biologist,**
public speaker, and Upper Current River District Interpreter,
Ozark National Scenic Riverways, U.S. National Park

A must read!! Robert Christie offers insightful and nuanced perspectives on challenges facing us as both individuals and as a society. A great dive into the complex social dynamics that keep us from facing and addressing existential threats.

~ **Zubi Wilson, Host of the radio talk show,**
"Living at the Edge," KSFR, Santa Fe, New Mexic

HOLDING
IT
TOGETHER

HOLDING
IT
TOGETHER

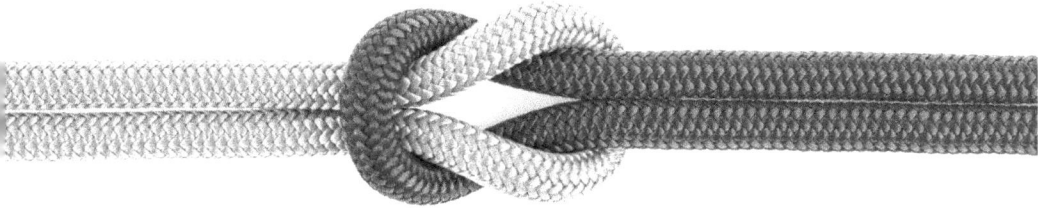

Social Control in an
Age of Great Transformation

ROBERT MACNEIL CHRISTIE, PHD

BEAR CLAW PRESS
SANTA FE

Bear Claw Press
An Imprint of Omega Publishers LLC
7 Avenida Vista Grande, B-7 #505
Santa Fe, NM 87508
USA

The Bear Claw Logo is a trademark of Omega Publishers, LLC
First published 2024

ISBN: 979-8-9907779-0-3 (paperback)
ISBN: 979-8-9907779-1-0 (eBook)
Library of Congress Control Number: 2024914908

Editing by Jen Marshall
Cover and Interior Design by Nick Zelinger, NZ Graphics

Names: Robert MacNeil Christie 1940-
Title: Holding It Together: Social Control in an Age of Great Transformation
Description:
Identifiers:
Subjects: LCSH: Social Ecology | Social Control | Consumer Culture |
Sustainability – Social Aspects | Climate Science – Social Aspects
| Social Policy – Scientific Aspects | Human Ecology | BISAC: SOCIAL SCIENCE

Bulk purchases of *Holding It Together* are available from the publisher.
For additional information contact: BearClawPress.com

AI DISCLAIMER: This book and all its contents have not received any input or influence of any kind from any form of Artificial Intelligence (AI) or similar system of content generation. Nor has the author engaged any AI system for research, assessment, or interpretation of any part of the text or of any materials from which the book has benefitted in relation to either facts or interpretations. This book is entirely the product of its author, the sole source of its content. In other words, I wrote it myself, aside from the published sources I cited.

First Edition | Printed in the United States of America

BOOKS BY ROBERT MACNEIL CHRISTIE

Hopeful Realism: A Climate Manifesto

Contents

Foreword

We don't face *environmental problems*. Ours is a *civilization challenge* on a scale unprecedented since humans left hunting-and-foraging for settled community. We could settle in because 11,700 years of climate stability, the late Holocene, allowed dependable habitats and social differentiation and structure. Now, amidst the stark climate volatility we've largely brought on, the early Anthropocene (the Age of the Human), we must negotiate a Great Transition, or a Great Turning, from Industrial Civilization to Ecological Civilization. This Great Work (Thomas Berry) is a collective calling that will span generations living amidst uncertainty.

Can we and future generations hold it together for this long, tumultuous season that tests human resilience, adaptability and creativity? If social control is possible, what kind is humane and fruitful?

Robert Christie's *Holding It Together: Social Control in an Age of Great Transformation* is the robust work of a social psychologist with more than a passing knowledge of the natural and physical sciences. For an era when it's all-hands-on-deck to do what needs to be done, this angle and its content are critical, critical for how we do things and why, with what consequences for life, all of it.

Christie names his vision "hopeful realism." It's unflinchingly honest about what is happening and why, giving no ground to wishful thinking and happy-ending illusions we might rather entertain. At the same time, Christie's realism underlines human responsibility, agency and choice, none of which is realistic apart from the human hope of meaningful outcomes.

Christie's hopeful realism is an alternative to stances most of us have felt and feared at one time or another.

There are the doom-sayers whose ranks seem to be growing. The young especially are fearful of the future, many of them toting a sense of impending doom that negates what should be their optimism and idealism.

Whatever the number of doom-sayers, they are still likely outnumbered by techno-industrial "green growth" optimists who remain convinced that capital and markets will find a path that continues rather than compromises unending growth and global consumerism. The grip of the extractive unlimited growth economy is strong, even when it turns destructive of the planet.

The more passive side of this optimism may be where the economically privileged majority reside. Their default is the live-in-the-moment, business-as-usual keeping of the ways we've long known, a status quo agreeable to us.

Last but not least are the masses of those simply set on survival, either because the social odds have long been against them or because they are now forced to join the company of the displaced, people and other species on the move because climate trauma has uprooted them. Whatever the reason, survivalist actions generally do not allow the long-haul time, perspectives and agency great transformations require.

Christie's invitation is a way through the mazes we encounter during the transformations already afoot. But it's only fair to ask about the *kind* of social control that holds it all together in this era of transition.

It's not the kind that tempts us as confusion, chaos and the nostalgia for the normal spread—authoritarian control embodied in a strong savior figure touting this as patriotic duty. Rather, it's democratic control tempered by hope and humility. That is, the kind of control committed to democratic processes and institutions precisely because these can lean into uncertainty and fear yet find ways to come together and work together. A commitment to processes like the rule of law and its institutions even as laws and those institutions change. Or incorporation of the rights of nature where those do not yet have a home in our processes and institutions.

This democratic control is humble and hopeful because it does not require a blueprint for some certain future, much less one for utopia. It requires the humility that we are not the some autonomous controlling species but one enmeshed in ecological interdependence that is not subject to human whim, desire and full-throttle control.

Finding a way here will, to be sure, tax our imagination to go beyond its home in industrial civilization. But fertile imagination, too, is best aroused and exercised by inclusive, democratic processes.

So can we hold it together? And if we can, will we? Frankly, I don't know; the future seems to specialize in surprise, especially when uncertainty is a sure thing for a long stretch. What I do know is this: That whatever happens, Robert Christie is one of the skilled, deeply informed, and trustworthy guides for what social control we might muster and manage. His is an invitation to accept, with gratitude.

Larry Rasmussen
Reinhold Niebuhr Professor of Social Ethics Emeritus
Union Theological Seminary, New York City
Author, most recently, of *The Planet You Inherit: Letters to My Grandchildren When Uncertainty's a Sure Thing* (Broadleaf Books, 2022). Winner of the 2023 Nautilus Gold Prize for best 2022 book in Ecology & Environment.
Website: larrywrites.info.

PREFACE

SOMETIMES IT SEEMS that there are three kinds of people in the modern world. The first, the techno-industrial optimists, believe technical innovation and the invention of new materials can and will fix anything and everything that poses a problem. The second, the modern doomsayers, the perpetual pessimists, remain convinced that no matter what we do, it will all come to nothing as the whole world implodes. The third group expresses quite explicitly that they do not care about the future of the planet or its people. These folks live for the moment and don't want to think about the future. I have met some of them and heard about others. They are the nihilists of modernity who refuse to take any responsibility for the human predicament to which they have contributed like the rest of us. They are not inclined to change their behavior. Each of these types has its theory of change—and they are all wrong.

But there is a fourth choice. Some of us recognize that we do have agency, yet at the same time cannot control everything—or even just what we want to change. I call that hopeful realism, the idea that to make any sense of it all we must be realistic and accept the best evidence available, no matter how disturbing, and work in that context until a better (more accurate) framing of reality comes along. Because we do have agency (we can make choices and act), we are at least capable of making some change in the world if not all that we would like—or all that we need.

When combining a realistic assessment of the world as it is, with the knowledge that we can act on that basis and possibly even make a difference, we have an honest foundation for hope. We cannot control everything, but we can (if we will) control ourselves and how we act in the world. By framing the world in that way, we establish a level and kind of hope grounded in reality—hopeful realism. However bad or good things appear, that approach is neither optimistic nor pessimistic. (Both are variants of fatalism, a futile and perhaps infantile perspective on life that is, which is actually kind of boring.)

When I began writing what would ultimately become this book, I focused on the many forms of social illusion that perplex and stifle humans in our everyday lives, personal relationships, and economic and political pursuits. However, bigger issues intervened in my thinking and my research. More and more, I considered the relationships between illusions and much larger (and growing) societal predicaments.

For a very long time, the destructive ways so many people and institutions treat our environment had concerned me. But when the pundits initially labeled the climate crisis by that soft, comfortable name, "global warming," I felt another kind of concern—the feeling that someone attempted to deceive me by intellectual sleight of hand. That was only one of the many ways by which groups and individuals would deny, downplay, denigrate as "a hoax," or dismiss the growing destabilization of the entire Earth System as something we could deal with in the future, if at all. Well, the future is now—and has been for quite a while.

Throughout the decades when "climate change," that other softball term, was studied, politically debated, and often dismissed, despite the growing objective evidence of the severe danger that growing disturbances in the climate system posed, I found the framing of the whole thing as "controversial" quite perplexing.

As a social psychologist, I felt fully aware of the human penchant for self-deception, in addicts for example, and the complexities of various belief systems as well. However, with my background in the natural and physical sciences, the trajectory of Earth System destabilization had become very real to me. You don't have to be an advanced climate scientist to observe the whole Earth System—climate, ocean, eco-systems, tropical forests, and more—begin to destabilize.

The only *real* controversies were a matter of refining our knowledge of how it was unfolding and what might be the most effective human response. (That is, issues of scientific substance around the timing and rapidity of the elements of climate and ecological destabilization, and over the relative importance of potential strategies of response.) These matters, of course, are not what the addicts of modern consumerism want to think about.

It will seem a wild radical viewpoint to some—well, maybe to many—when I argue that we are rapidly approaching the end of industrial civilization. The idea is entirely outside the dominant worldview of techno-industrial modernity. The modernist worldview holds "progress" to be inevitable and permanent, due mostly to tech-nological innovation and unending economic growth. The consumerist propaganda is so intense and pervasive that hardly anyone considers an alternative view.

But the real-world trends that scientists and others have documented for decades tell a very different story. The basic science is clear. We live on a finite planet and we, the rapidly growing human population, have overshot the capacity of the Earth System to carry the load of our extraction, production, consumption, and waste. Most of the responsibility rests with the former colonial nations of the Global North that plundered the world's resources and people, especially in the Global South, since the earliest adventurers of the "Age of Exploration"—and continue to do so.

Donella Meadows and her MIT colleagues accurately predicted in 1972—over four decades ago—that Nature's materials and systems, which feed the industrial machine, would begin depleting right about now. (They were widely ridiculed, and their research was dismissed by all of the "most important" economists at the time.) We now know not only that their report, "The Limits to Growth," was right on target. We also know that there is much more involved than that.

To understand the trajectory of industrial civilization and imagine where we might go from here, we must look at the growing indicators of instability at every level from local ecosystems, towns, factories, and farms—to global climate and the ramifications of changes in weather patterns that farmers around the world had depended on for thousands of years. On top of that, the more recent global political economy financialization, amid increasingly frequent wars and other disasters, does not auger well for the dominant ideology of endless economic growth. The illusion that somehow our destabilizing, finite planet can survive much longer without spiraling down into deep chaos under the relentless pressure of growing human-caused disruption seems more tenuous than ever.

Can we hold it together? To be clear, I don't mean that in the sense that most defenders of the status quo see it. To them, I must answer with an unequivocal "No." The end of industrial civilization will occur in one of two ways, only one of which suggests that we have held it together. It may very well collapse if we fail to deploy our natural human capacity for resilience and make the radical changes in our lives that we must. Instead, we may transform or replace it with a set of diverse ecological societies and communities integrated with the local/regional ecosystems of which they are a part, no longer separated from Nature. We can and must become a newly responsible part of our ecological communities.

This book explores how we got into this global predicament, the key elements involved in where we stand now, and what looks like the best way out of this mess. Can we hold it together enough to step outside our long-held illusions and take radical new actions to address our radical new world?

CHAPTER ONE

Normal, or Not

FEW PEOPLE WOULD disagree with the assertion that we live in a world of increasing uncertainty and danger. Repeatedly over recent decades, surveys have revealed the consistent opinion by respondents in the US that the nation is heading "in the wrong direction." Unfortunately, such surveys fail to dig deeper and explore the idea of the direction of a nation—or what exactly is wrong. As a young adult in the 1960s, I felt that the nation was tragically misdirected. Perhaps that is the source of my lifelong interest in social change, which eventually led to writing this book.

Ever since the 1960s, it seems that the richest, most powerful nation in the world has experienced one crisis after another. Or maybe discontent is universal. However, certain themes have emerged as dominant, mostly because they represent unresolved problems of purpose and direction, fueled by unremitting change. The continuous failure to get at the root of problems seems buried in the perpetuation of certain social illusions, not the least of which some call "American exceptionalism." In several ways, the US *is* an exception among nations, though not necessarily in the glorious ways implied by that trope.

It is a long list, although we could also conclude that Americans are not really so different from anyone else. It is just as easy to argue for many key universal traits of humans, as for national or ethnic differences. When we look to the root causes of problems, purged of ideological biases, differences of opinion sometimes resolve into consensus about goals, which too often fail to influence politics. I wrote this book from the perspective of an American who has experienced well over a half-century of the industrial-consumer modernism of a nation that has dominated much of the world by its sheer economic, military, and political power through most of the twentieth century and into the twenty-first. The implications of this situation are elemental for the future of the planet, and us.

For me and a growing number of people I call outlier economists, along with a small number of social critics and most climate and ecological scientists, the defining feature of this era is the headlong rush into catastrophic change. That fatal flaw is inherent in the perpetuation of the myth of endless economic growth on a finite planet. The much-touted process of globalization is the engine that has led us to an unprecedented planetary predicament. (Ironically, the frequent use of the word "unprecedented" has itself become unprecedented in recent years.) An overwhelming body of evidence suggests that this trajectory of the global political economy is terminal.

For years, I had intended to write a book that I thought I might call *The Social Illusion* because the misapprehension of reality causes so much of what goes wrong in the world. Social illusions result from political, economic, and social ideologies and the powerful interests that drive them. The trend of increasingly severe crises grew worse because decision-makers failed to replace factually erroneous economic theory with reasoned responses to the evidence of growing crises. The facts of science applied to the goal of realizing core human values, it seemed to me, demonstrated that failure. I pursued those facts for many years but never got around to writing that book.

Retiring from teaching as a professor of sociology at California State University allowed me to put enough time into the writing and research that inevitably took the project in a somewhat different direction. The role of social illusion remains central to my understanding of societal misdirection and global predicaments (resulting in large part from the power of economic and political elites). However, the growing complexities of interconnected global crises turned my attention toward the relationships between modern human societies and the growing disturbances to our home, the Earth System.

These very changes, which occurred in the last few decades, forced me to shift my focus to the global crisis of climate-ecological destabilization. The industrial-consumer ideology and its political-economic imperative of endless growth on a finite planet are central to the crises. What I now see as the greatest predicament that has ever confronted humanity has directly resulted from the entirety of Western civilization. The Industrial Revolution and its technological and organizational elements, which have played out in the last two hundred years of the Industrial Era, are at the root of the existential threat to planetary stability and human life.

Hardly anyone talks about it, but we are fast approaching the end of the Industrial Era, which involves a global process I call the New Great Transformation of Industrial Civilization. The outcome of that transformation is as open a question as our capacity to respond quickly and effectively to the self-inflicted danger facing us.

Let me be clear. We face two New Great Transformations occurring together. One is the destabilization of the whole Earth System, and the other is the destabilization of the global political-economic regime. Industrial Era impacts caused both, and both now require extreme human intervention to have a chance of a tolerable future. One of the biggest problems is that of perception. In so many places, things seem quite normal, both economically and environmentally. But they are not. Some effects of change may not show up right away. But when they do, it is clear that they are not normal.

The Earth System, our home, is already experiencing a New Great Transformation with the destabilization of its subsystems: the atmosphere (air), lithosphere (land), hydrosphere (water), and biosphere (living systems). Emerging catastrophic changes are well underway in each of these systems. The effluents of the Industrial Era have directly caused the decline of the many elemental systems in Nature upon which we depend for our survival.

The damage has increased severely since the advent of fossil-fueled energy systems as the driving force of modern industrial economies. It continues unabated, and it has accelerated right up to the day you read this. That disruption of Nature already affects human populations as the juggernaut of economic growth and its carbon emissions persist. Even more importantly, the disruption of natural systems is accelerating at an alarming rate, especially in the last few decades.

Complex adaptive systems like ecological systems, human societies, and the Earth System itself cannot easily predict changes resulting from disruptions because they are nonlinear. The complexity of nonlinear systems results in uncertainty about the exact consequences of any particular change. Simple linear cause-effect models of change do not apply where outside interventions disrupt multiple interacting complex systems and their subsystems. Nevertheless, the broad outlines of change over time for the climate and for ecosystems are relatively clear.

We know—as Exxon-Mobil scientists knew for decades—that the injection of large quantities of CO_2 and other greenhouse gasses into the atmosphere causes it to warm, disrupting climate worldwide. Climate is a central factor in all ecosystems. However, it is difficult to predict accurately the exact consequences of such changes in specific locations at particular times. Destabilization due to heating may lead to oscillation, as the effects of a change in one small component may spill out across the larger climate system, which, in turn, would alter conditions in other systems.

We cannot predict the occurrence or exact path of a hurricane, for example, but we know that with the warming of the ocean and the atmosphere over time, hurricanes that occur in the Caribbean are likely to be stronger, drawing additional energy from the warmer seas. Oscillations of the El Niño and La Niña patterns of temperature change in the eastern and western Pacific Ocean, and many other factors, may be involved. The bottom line, however, is that what had been relatively stable cycles and patterns over tens of thousands of years are becoming increasingly erratic as the atmosphere, ocean, and land heat up. Such instabilities affect humans and vast numbers of other living systems all over the world.

We industrialized humans have caused the Earth System to destabilize rapidly, and too many have denied it for too long. Once recognized, both public and private institutional "authorities" have consistently failed to respond in any rational way for decades. They have repeatedly made empty promises to cut back on carbon emissions and restore ecosystems or to keep average global temperatures below 1.5° or 2.0° Celsius above pre-industrial levels by this or that date. Greta Thunberg expressed the problem quite succinctly in 2021: "Blah, blah, blah."

The science of what institutions and people must do to stop the damage and restore the Earth System to a balance is clear once we purge the political-economic equivocating from the discussion. Above all, we must reduce emissions from the burning of fossil fuels to near zero very soon. As I write, they continue to rise despite that knowledge while the false promises continue. Such massive change cannot happen within the framework of existing economies without radically changing the scale and kind of energy used by modern economies. That means restructuring economies to use much less energy and to change production processes, even with a transition to clean renewable energy production. That also means radically changing the way we live in modern societies.

However, few officials even consider, much less understand, how to accomplish the complex societal changes required to do what is technologically necessary. They apparently do not recognize any societal implications of making these massive technical changes. The disturbance of human systems, all of which are complex and adaptive in diverse ways, has consequences as difficult to predict as the local effects of a changing climate. Serious public discussion of the complex societal changes needed to achieve near-zero carbon emissions in hopes of restabilizing climate and ecological systems has yet to occur.

Unfortunately, we have done very little to explore the broad outlines of societal changes that result from climate disruption. We also have not considered the actions we must take to adapt our ways of living to rapidly changing conditions moving into the Anthropocene—the geological epoch in which human activity has become a major factor affecting the Earth System. However, human systems have one unique feature. Their components—persons, groups, and institutions—have a level of agency far more powerful than that of any other creature.

That is the source of our hope.

Humans can make rational decisions and do things, or not. That is, via consciousness and culture, we can organize ourselves to take intentional action as persons or groups to change the world in which we live. Throughout human evolution, we have taken individual and group action in the interests of our survival, comfort, fun, or simple greed. Today, the actions we must take are unique in human history. They are extremely complex, and we do not yet understand them all.

The complexities and consequences of the societies of the Global North isolating our lives from Nature, both materially and culturally during the Industrial Era, are the subject of this book. Given the trends we now see, we have no more than a decade or so to get it together and take lines of action that will be extremely difficult but imperative. We must realize that our survival depends on initiating a New Great Transformation of human societies themselves to form an ecological civilization.

The biggest question of the first third of the twenty-first century is if we can radically change the way we live by holding it together enough to make the changes that few have even contemplated before now. I could never have imagined, as I enjoyed the privileges of growing up as a working-class kid in what now seems the Halcion days of blissful tranquility living the good life in mid-twentieth-century Southern California, that we might ever arrive at where we are today.

LYING ON MY BACK in the warmth of my cozy flannel-lined Boy Scout sleeping bag, gazing at the dense display of stars above the high desert, was an intensely curious, yet comforting, experience for a ten year old. It was not very cold, but we could feel the temperature dropping fast during the hour after sunset in the high Mojave Desert. We were camping north of the San Gabriel Mountains in Southern California at an elevation of about 2,800 feet in 1950, well clear of the smog-laden Los Angeles basin. The sky was pitch black and clear, the stars were extremely bright, and shooting stars were out in full force.

The wonder of that experience was so breathtaking that I could not fall asleep until finally overcome by the ultimate peacefulness of it all. The jabbering of us boys debating constellation names and locations in the sky gradually faded into an intense desert silence. We had learned the names of most of the constellations and tried to identify them while waiting for the next meteorite to flash across the heavens. The waits were brief. It was such a powerful experience that I remember it clearly, even today. The world and especially the universe were a great mysteries, yet, it seemed to make sense somehow. It felt knowable and right.

The little we had learned about astronomy, however, did not diminish the infinite wonder of the night sky in the high desert. Why did the Greeks see gods in patterns of stars in the sky? Where did those meteorites come from? The stars yielded no answers as we dosed off. Now, I look back across the decades with an entirely new set of mysteries on my mind.

To a kid, the difference between reality and illusion exists mainly in magic tricks. As you get older, if you keep looking with a critical eye, things get much more complicated. If not, they may remain simple and certain, although not necessarily burdened by understanding.

For me, the strange relations between illusion and reality seemed more important than ever as I moved through various social worlds and times. Over the decades, the edge of illusion has become ever more complex and potentially dangerous. How is modern industrial civilization even possible amid the growing chaos and danger surrounding us? Well, maybe it no longer is. The times we live in are almost too interesting. So much seems to be going so very wrong. Many of the most basic questions remain only partially answered, if at all.

Back in my youth, I knew almost nothing of history, the social order around me, or societal change. I only knew my family, neighborhood, friends, and the thrill of the moment as I discovered new things about the world of "now." As I anticipated new adventures, I felt impatient to discover more. Even as I marveled at the order in the world, I gradually came to realize that not everything was right in the larger world around me.

Even in the early 1950s, the Los Angeles smog was so thick on some days that it was difficult to see across our schoolyard below the foothills of the San Gabriel Mountains, which kept that toxic soup from escaping the Los Angeles Basin into the Mojave Desert to the north. It made me wheeze and cough. I wondered why smog intruded into such a beautiful world where the scent of orange groves amid the scattered suburbs always pleased me as I passed on my bike rides along the foothills.

But, to a child, the whole world is normal. Children have no other point of reference. Even the abused child, having seen only abuse, assumes its normality until it becomes unbearable. By the time we reach adulthood, the world we know is normal by default until our experience expands. Leaving that child's world, however, can make all the difference. Leaving home is inherently to leave the normal for the unknown. Coming home again, normal becomes an open question; nothing is quite the same.

I had no idea then how complex and uncertain life would become as I sought answers to questions about the future, most of which I look back on now. I could not have imagined that eventually I would live through at least the beginnings of the greatest transformation of humanity ever. It would be at least a decade before I would turn my attention to the increasingly obvious cracks in the social and natural order—and wonder what had gone wrong.

The Golden Age of Social Control

We are familiar with the era in American history following shortly after the Second World War for its rapid economic expansion, which McNeill and Engel (2016) called The Great Acceleration. The prosecution of the war had built up a huge industrial capacity and a lot of pent-up consumer demand after rationing certain materials, even some foods. The wartime constraints on the supply of goods had ended, and a great capacity to supply energy and materials to drive economic growth had resulted from the unified war effort. The tremendous relief that the war had ended provided motivation and released civilian ambitions to get on with normal life and seek deferred dreams. The fossil-fuel capacity that had energized the war effort, suddenly available in abundance, lay waiting to drive economic growth.

Righteousness had defeated evil, and the American ethos expressed a great optimism for the nation's future. I was five years old when the war ended and The Great Acceleration began the golden age of American expansion and prosperity. A high level of social consensus prevailed, opportunities flourished, and hopes were high. I experienced the rest of my youth in what you might call, looking back, the era of great opportunity to succeed within the existing social order.

However, opportunities did not occur evenly. As I approached adulthood, I gradually learned that opportunities were highly con-strained for Black folks, Mexican Americans, Native Americans, and other peoples of color, as well as poor (mostly rural) white folks. This was especially true in the South due to an inability to benefit from union wages. In the 1950s, union membership had peaked, and many urban workers enjoyed middle-class wages.

It was not until the late 1960s that the cracks in the system of social control appeared increasingly obvious. My academic studies comple-mented what I saw directly in the society around me. I had gotten a taste of racial discrimination and prejudice while in the Air Force. The military, formally integrated, constrained blatant personal prejudices, at least on the surface. In the winter of 1963, while stationed in Orlando, Florida, the Air Force granted me a leave to drive back to California for Christmas with family. They inadvertently offered me the experience of the deep Southern culture of racial discrimination.

Lamar, one of the guys I worked with in my geodetic surveying unit, was from Tyler, a small town in the middle of East Texas. Lamar had mentioned that he planned to go home for Christmas too. I offered him a ride. I had just bought a brand-new MG Midget sports car with nearly all my savings—just under a thousand dollars. We could share driving and the cost of gas until we reached Tyler where I would drop him off and continue west. It would be a win/win situation, with an experiential bonus I had not expected.

Lamar was Black. When I offered him a ride from our base in Orlando to Tyler, he was hesitant. Of course, I was aware of racism in the South but only indirectly. Throughout high school in Southern California, I had worked summers on construction crews alongside Black men who were my construction mentors and I considered friends, even though they sometimes teased me over my unprofessional management of a shovel. I occasionally overheard some White workers verbal expressions of prejudice. One might say I was naïve at age fifteen.

Lamar informed me of some of the things we would have to do to drive in relative safety—as a White man and a Black man together in a sports car—through northern Florida, southern Georgia, Alabama, Louisiana, and nearly halfway through eastern Texas. To eat, Lamar explained to this naïve Californian, I would have to go into a restaurant at the main entrance alone. He would go around to the back to buy his food separately. That is exactly what we did.

I felt very uncomfortable, maybe even a little guilty, as a person "who thought he was White." (Baldwin, *The Fire Next Time*.) I had rediscovered that phrase much later reading Ta-Nehisi Coates' book in the form of a letter to his teenage son on the perils of a Black boy growing up in America at the turn of the twenty-first century; I had long since forgotten Baldwin's use of the term, but when Coates used it, it struck a deep chord. The point, of course, is that race is a social construction, not any kind of biological category, but as a social meaning, it is very real.

The fact that a person thinks he is White inevitably bears on his consciousness and behavior in relation to any person who might at other times have been called a nigger, nigra, colored, Negro, or Black—or any of a long list of pejorative terms. So, in that context, across the street from a restaurant in a small Louisiana town in 1961, I could not have felt anything but uncomfortable, as well as outraged. I knew that I was in a situation that was terribly wrong and that I had absolutely no possibility to do anything other than go along with it if I wanted to get Lamar and myself home for Christmas.

There would be no sit-down meals at a lunch counter on that trip if we were to expect to continue our journey unharmed; it was 1961 in the American South. (I had occasionally eaten lunch in diners in Los Angeles with my Black co-workers in the late 1950s, though we usually brought our lunch to work.) The direct experience of my friend's oppression was profoundly depressing.

At night, I would have to drop Lamar off at a railroad crossing so he could find lodging, literally on the other side of the tracks. Then I would find a motel for the night near the highway and meet him at the tracks in the morning. No smartphones: we just set a time to meet. The whole thing was humiliating—and rather scary. I thought a lot about how it would be to live like that all the time.

There are several forms of social control. One, of course, is through intimidation or threat of physical violence by a dominant group. Another is actual violence. The Jim Crow South exemplified both, with many remnants remaining today in slightly different forms, even after the formal gains of the civil rights movement. The rise of neo-Nazis and White nationalists in the first quarter of the twenty-first century, like the rescission of the voter rights law by the Supreme Court, confirmed the depth and persistence of racism in America.

Persuasion is another means of social control where the implied or specific threat to the target of control signifies that they will lose something of value by not complying, or will receive a reward for obedience, or even by logic without direct consequences. On the other hand, self-governing communities use mutual consent to produce social control. People hold similar values and cooperate on mutual goals because they trust each other. In communities of mutual respect, persuasion becomes part of the process of reaching a consensus, rather than being a means of manipulation, as it is in the consumer culture.

I recently heard of a pastoral tribe in northern Iran whose culture does not allow for leaders. They simply negotiate conflicts to resolution without Ayatollahs involvement. In hunter-gatherer groups, social control results from members embodying different complementary roles in provisioning the group with food and other basic resources. Consensus about identity and responsibility is clear and is the essence of social control.

My experience of the oppression of Black folks in the Deep South in the early 1960s was probably the most powerful direct exposure to social control gone wrong—that is, violent enforcement of a hierarchy of oppression—that I have ever had. It was certainly the most memorable. It informed the rest of my academic career and much more, especially my outlook on the many examples of oppression I have since observed. This form of social control always rests on the dehumanization of the oppressed by the dominators.

Despite such cultural abominations as the White supremacist control of Black folks in the Deep South, in post-World War II America, the White majority experienced expanding affluence and a relatively high degree of cultural consensus. Almost everyone agreed with the generic goals of individuals taking opportunities to build an economic future for themselves and their families.

In effect, most people agreed on a set of basic norms of behavior. However, little tolerance existed for any form of deviance from social norms. There were no such liberties such as gay rights or even tolerance for unusual expressions of personal style. It was clear what society considered normal or abnormal. The term, LGBTQQIP2SA did not yet exist. We can certainly criticize aspects of the conformist culture of that time. By the mid-1960s, the youth of America did just that. Nevertheless, deviance from norms could engender social rejection—and even violence.

The era of normative conformity amid expanding opportunity shows how societies can work for many when consensus is high and conflict is low—unlike the America we are experiencing now. That all began to change with the emergence of the "counterculture" and protest movements against the Vietnam War and civil rights violations in the late 1960s, even as economic acceleration continued into the 1970s.

Once that era faded, the economic and political elites consolidated power by suppressing the democratic expression of the public interest. The new forms of hierarchical social control, social disorganization, and personal problems became more severe. To paraphrase what C. Wright Mills said so presciently in 1959, personal troubles align closely with public issues.

The Tenuous Order of Things

IN MY YOUTH, when I saw things that were not right, such as discrimination against Black folks or the smog-laden LA skies, I looked for causes. I did not find very satisfactory ones. Understanding how smog happens technically is one thing, but figuring out how to stop its production is quite another. Maybe that triggered my lifelong interest in social change. Calling a racist a racist may be accurate, but it does not explain racism.

Smog, for example, results from the actions of a whole lot of people and the machines they operate, and generally, a faraway corporation manufactures most of these machines. However, its connection to the growing number of cars driving all around LA was indirect or absent in people's minds. A whole complex of social relations and manufacturing design and operational deployment is to blame. That is a big part of why fixing the problem has been so difficult. On top of that, a whole complex set of industries and their owners have every interest in avoiding an expensive fix. They exercise that interest in the political realm to minimize their responsibility for change. That is a major factor, of course, in the denial, obfuscation, and resistance to dealing with the growing climate-ecological emergency today.

When we think of social control, it is essential to consider who controls what—and to what end. Whatever hierarchy of control exists in a society, it involves a certain degree of control over the channels of communication that shape whatever consensus exists. We tend to believe that we have created our own thoughts and beliefs, but we mostly hold the ideas and beliefs of our culture or subculture—and we express them through our behavior, which is very much like that of our peers. The existing channels of communication sustain beliefs and norms, and the relations of people within their social networks reinforce them.

We call the results social structure. This element usually involves a significant degree of hierarchy, especially in industrial-consumer societies, and of course, in empires of various sorts. Hierarchy exists not only in terms of wealth and poverty. It also appears in social position, status, and in our perception and valuation of different groups. The notion that might makes right is always implicit in a hierarchy. At the same time, we imagine ourselves to be independent individuals, sustained as such by the democratic processes we have failed to maintain.

The rest of this chapter is all about how social structure works for—and against—people and how tenuous it really is. Within industrial-consumer societies, we tend to hold it together in "normal" times. Most folks live in a relatively stable relationship of dependence on a stable existing economic structure. However, we no longer live in normal times. Dependency prevails, but stability and comfort? Not so much.

Holding It Together in "Normal" Times, or Not

Sociologists refer to the process by which individuals grow up amid family, neighbors, and communities and internalize the behavioral norms, as well as the values and beliefs that prevail in their society, as socialization. When people comply with norms because they have internalized them and treat them as their own, self-control becomes social control. There is little if anything to enforce. Self-control is the strongest form of social control, resulting in system stability. That happens when a high degree of consensus on values and beliefs prevails.

Of course, reality is never that simple, clear-cut, or convenient, especially in multicultural industrial-consumer societies. Language and meaning are always subject to individual interpretation and often distortion in service to selfish desires or special interests. Depending on many things, that may result in "deviance" from norms or violation of laws. Beyond that, sometimes serious conflicts arise over what behaviors society accepts—or not. Disagreements arise when people with different personal or institutional interests seek to establish their norms or policies as the rule for everyone else. These conflicts often stem from opposing ideas about what is best for everybody.

In the United States of America, in the 1950s and early 1960s, the vast majority of Americans agreed on most norms and values. Of course, there were struggles over economic power and social status, and there was certainly growing prominence of "social deviance," especially among the young. However, stability resulting from normative consensus was the dominant theme in the larger national culture. Back then, "conservatives" were conservative and held no ideas of insurrection or even imagined overthrowing an election.

To be conservative meant that one held to the norms of civility and strongly believed in upholding the nation's laws and its constitution. Liberals and conservatives enjoyed civil debates over policy and the norms of civil society, within a relatively narrow band of what most folks considered debatable. We viewed policy outcomes as the legitimate result of a democratic process, even as elites conspired to exert control to the extent that their wealth and power might allow.

Today, far-out conspiracy theories become excuses for all-out violations of norms of civility and political order, even to the point of inciting violence, crossing all political boundaries of civility and tolerance. Disaffected people, for whom fascist ideas are increasingly attractive, violate traditional norms of civil behavior. They lean toward political violence, as affiliated politicians support such behavior as acceptable and normal. W.B. Yeats expressed this kind of political disorder well in an earlier time of great turmoil, 1919, in the wake of World War I.

Things fall apart; the centre cannot hold;
Mere anarchy is loosed upon the world,
The blood-dimmed tide is loosed, and everywhere
The ceremony of innocence is drowned;
The best lack all conviction, while the worst
Are full of passionate intensity.

The close fit of Yeats' words with conditions today is nothing short of astounding. It would seem that the center cannot hold, as massive Earth System changes call upon us to work in diligent concert for our collective survival. Yet, as we head for the big-box store, our institutions do little more than offer empty gestures. Can we hold it together long enough and with enough fortitude to make the transition away from a self-destructive, fossil-fueled, global-corporate empire of endless growth on a finite planet? Can we achieve an ecological civilization that lives in harmony with its habitat, the living Earth System? As the saying goes, the jury is out.

The potential for human flourishing in a rapidly changing world remains an open question. And given the accelerating destabilization of the Earth System and human systems as well, the prospects are not good. We need a New Great Awakening to acknowledge that the transformation of climate and ecosystems, so vital for human survival, requires a level of institutional change that few have considered. That can happen only in the context of a new vision of the human future.

Order and Chaos

In everyday civilian life, simple ideas of "law and order" and "law enforcement" imply that most ordinary people comply with static behavioral norms, institutional rules, and formal laws—and some others do not. That may even extend to beliefs about social conformity and unacceptable lifestyles. Most people behave "normally," at least in public, but since the right to privacy is a core value, norms of privacy also tend to prevail.

Increasingly, however, expression of lifestyles that deviates from norms—purple hair, tattoos, sexual practices, evolving gender identities, etc.—have gone public. And many seemed to have tolerated and even accepted them until recently. The rise of political extremes and the re-emergence of hatreds, such as White nationalism, antisemitism, and xenophobia includes the suppression of lifestyle choices at odds with traditionally dominant cultural norms.

Simplistic visions of "good guys" and "bad guys" now seem to apply to almost everyone, with little room for subtlety in between. Social pressure and a rigid belief that "they" are evil seems to justify the evermore ruthless enforcement of political ideology and social prejudice. That sometimes applies to the enforcement of laws too. Nevertheless, law enforcement often experiences pragmatics that are just as messy (nonlinear), although rarely as bloody as military combat. At the same time, we should be acutely aware that the rise of fascism has always depended on the dehumanization of a subjugated group (Stanley 2020).

Laws are imperfect and subject to diverse interpretations, inconsistent enforcement, and legislative change. That is why sustaining an impartial non-partisan Supreme Court is so important, despite its recent failures. Impartiality declined once Trump stacked the court with young, inexperienced authoritarians approved by the extremely conservative, or rather reactionary, Federalist Society. The degradation of judicial non-partisanship was a very bad sign for the survival of democracy. It offers cultural and even legal support for autocratic attempts. Any complex system requires a certain level of order and can tolerate only so much chaos before becoming unstable and subject to collapse. Genuine social order results from cultural consensus, not political alienation and civil strife.

Legislation distributes the benefits of "law and order" rather unevenly since the most powerful interest groups write or heavily influence the majority of legislation. Some laws apply only to specific economic situations, such as finance or commerce, so they seem irrelevant to the everyday lives of most citizens. Many other laws are far too complex for any one person to fully internalize or avoid violating in some way. Creativity fills in the gaps. A police officer following you for a few blocks can always find some technical violation if they want to. That now has a name, the "pretext traffic stop." At the same time, most laws do not even pertain to everyday life or the behavior of the average citizen. They refer to some arcane aspect of institutional operations, bureaucratic procedure, or political-economic authority.

However, generally, police operations focus on public behavior, which is the easiest to observe. That is one of the reasons, along with class prejudice and racism, that the poor and people of color suffer the effects of frequent arrests and convictions. (They usually plea-bargain down from a more extreme charge to a lesser felony, without the benefit of a trial before a jury of their peers). This stigmatizes them for life as a felon, making it very difficult to obtain employment. White-collar crime happens in corporate boardrooms, accounting and legal firms, and, above all, in private. Police ordinarily become involved only after someone exposes the crooked deal.

People behave mostly in the ways they learned growing up in their families and their neighborhoods. Moral and legal principles are far less exact and clear-cut than many other complex practices, such as the Blue Angels performing disciplined procedures and practice of high-speed aerobatics or Simone Biles' precision gymnastics. High-precision athletics involve many years of practice and integration of refined behavior into the identity of the individual, which embodies very precise practices.

The details of the legal system integrate much more loosely with individual personality, if at all. All sorts of political, economic, organizational, and personal interests infiltrate the ethical and legal principles that we assume guide or fail to guide behavior in public and private everyday life. Personal moral codes, whether strong or weak, result from life experience, not law.

Interests, experiences, and habits shape interpretation; interpretation governs both compliance and enforcement. Social loss by particular groups may produce interpretations that posit the clandestine workings of imagined conspiracies that must have caused, for example, the loss of economic and social standing of White working-class men in recent decades. Demagogues exploit the fears and anxieties that accompany widespread loss of status and income to produce hatred of vulnerable groups, such as immigrants and ethnic groups. This foments hate crimes and political violence, resulting in even more social instability. When status anxiety and economic losses grow in a society, demagogues gain greater potential to take power in that society.

Most folks operate on what they know to be good or bad, right or wrong, or what they think they can get away with. The language of human behavior is rife with opportunities for rationalizing a particular behavior by an application of high-sounding principles. Parents and teachers attempt to instill morality and ethics in their practice of socializing children. Yet, social control is fluid, differentially internalized, and not entirely predictable. Alienated groups, especially when encouraged by authoritarian leaders, may attempt to exert social control in contravention of law and political ethics by fomenting and participating in political violence.

In a theocratic state—an extremely hierarchical political system—those in power may apply force to control everyday behavior, such as in contemporary Iran, where so-called "morality police" brutalize women who fail to cover their hair with a hijab in exactly the prescribed way. In an ostensibly democratic nation where levels of status anxiety and economic insecurity grow high, authoritarian non-state actors, such as the Oath Keepers, Proud Boys, and various White nationalist groups may use violence against both the groups they fear will "replace" them and the political authorities they view as traitors for not supporting their extremist views. Such groups also typically espouse male dominance (patriarchy) and racist and authoritarian ideas.

The term "socialization" refers to the integration of individuals into a culture. Parenting, the primary mechanism of early socialization, varies a lot in both intensity and effectiveness. Today marketing increasingly overrides parental authority with images of self-indulgence, rebellion, and power, most of which aim to encourage childhood consumption rather than independent thinking or adoption of basic cultural values.

Individuals, therefore, grow up with varying degrees of self-control and skill in managing and justifying their behavior under changing conditions. Many even grow up with identities more attached to consumer culture, applied through peer pressure and marketing, than to their families and communities. When identity vests with ideas from consumer culture rather than natural social institutions like family and community, the individual tends to become alienated from the underlying social order—such as it is. The institutional milieu we all must engage in at some level tends to force the social self into rational-legal categorizations that isolate self-image from membership in family, community, or social group. Core human values not part of that larger techno-industrial matrix that Dmitri Orlov (2017) calls the technosphere fade into the background for many. Modern life in "advanced" societies reminds me of the science fiction movie *The Matrix,* in which Keanu Reeves's character, Neo, moves between illusions of both power and disaster while trying to distinguish between reality and the surreal matrix into which he resists seduction.

Unfortunately, we may know less about the future we face with a shrunken technosphere at the end of the Industrial Era than we imagine about the world Neo faced beyond The Matrix. Just beyond the tenuous social order is the looming prospect of chaos and collapse. The question of the century is if we can shape a new societal order in harmony with the laws of Nature and the principles of human evolution that carried us through most of the human journey.

The power of science, technology, and social hierarchy in the Industrial Era has sent us down what is now recognizable as a terminal path. The following chapters explore the obstacles and opportunities inherent in humanity's greatest predicament ever: our present condition and trajectory. It is extremely difficult to forge an entirely different path in this context.

CHAPTER TWO

Social Control

SOCIAL CONTROL IS not only the central process by which societies function. It is increasingly problematic as we approach the end of the Industrial Era. Yet, it is a very difficult thing to discuss. For most Americans, it is an uncomfortable concept to contemplate because our culture values individual freedom (self-control) unequivocally—as long as you behave as expected by those who hold power. Social control seems the opposite of personal freedom. Nobody wants to think that someone else or some unknown entity has the power to control their life. Yet, it is only by virtue of social control that we can exercise freedom. The absence of social control is social chaos.

We tout our personal freedom, yet we routinely succumb to social pressure or embrace formal authority. So, why should we focus on social control? Well, real-world freedom is never absolute; it exists because of our place in society and the opportunities that affords us— or not. Consider the escalating political resentment and chaos in the industrially developed nations of the Global North today. This trend mirrors similar situations in history, most of which led to the loss of democracy, political tyranny, and bloodshed—that is, the loss of real freedom. Changes in social control are critical for understanding and making a difference in the trajectory of our lives, not only in the future but now.

I think it is important to look into how social control works to better understand the actual sources of people's behavior, especially the coordination within groups and organizations, ("the social order," such as it is). We need our personal freedom, but to have it we also must coordinate with one another, often in quite complicated ways, just to create and live in a civilization where personal freedom is possible.

The kind of social control that exists in a particular society matters for both personal freedom and the quality of life we experience. Social control can go off the rails, producing severe instability or worse. Anti-democratic forces can result in tyranny and ecological chaos. Let us be real: humans are social animals, and we associate and coordinate with one another naturally all the time. Our freedom emerges from how we conduct our social relations, which always entails some level of constraint. However, the more complicated things get, the more likely it seems that the systems we build become vulnerable to chaos and collapse.

Who and what actually controls how both our organizations and our selves behave? Who controls the social order and how? How well is power shared? Well, that is a far more complicated question than it appears at first. Look at politics for example. In election campaigns, politicians routinely blame their opponents for what is not right and take credit for whatever goes well. Yet, if one examines the facts regarding inflation, unemployment, crime, etc., the conditions during the first two years of any presidency mostly result from previous events and policies enacted before the current president assumed office. The party out of power blames the current president for anything that may have gone wrong, then typically wins more seats in the House and Senate in the midterm elections.

Multiple people and institutions within a nation control societal conditions through a complex interplay of prior policies, actions, inactions, and ongoing relationships and by external states and non-state actors as well. It can take years before anyone feels the results of some policy changes. Yet, politicians use the immediate situation or merely conjure a false narrative to blame others or take credit daily.

To understand all this better, we must get down to the basics of social organization and personal behavior. Neither the evening news nor social media are up to that task. Most people think of the "social order" in terms of the law, rules, and regulations in organizations— and conformity with or deviance from various rules. However, we must look at history, where we will find many clues, and we must look at the findings of research in psychology, sociology, political science, and economics.

But wait, there is so much more. How do we know what we know, and what makes us do what we do? Folks say that they make their own decisions and behave accordingly based on their values and beliefs. However, it is much more complicated than that. We all live in the context of the lives and behaviors of others. We constantly influence each other, whether we are aware of it or not. Social control takes diverse forms; we must understand how each works and for whom—for everyone or just for the powerful? The implications are huge.

How We Know Things

When we are young, everything is interesting. With time and the pressures of daily life, however, we may lose interest and wonder. We may settle instead for the security of routines assigned to us by bosses or accepted as social conventions. We adapt our beliefs and our behavior to the conditions we live in. We are creatures of habit, much more so than we believe.

Sometimes a larger reality forces itself upon us by challenging our most basic assumptions in such a way that we feel wonder again—or, more often, just fear. Some people strongly resist facing new realities. We all seek stability to some extent. Fear and curiosity, along with habit, determine the extent to which we deny or seek to understand new facts. If fear is stronger, we may retreat into the security of our group belief systems, which may ignore or deny the new reality. We may instead descend into a belief in conspiracy theories to give meaning to our fears.

If curiosity overcomes fear, we are more likely to question what in the world is going on, especially if we sense the growing chaos around us. One person's reality is another's illusion. The whole issue of how we can distinguish illusion from reality has always fascinated me. I am now convinced that fear and habit strongly influence our worldview and our misapprehensions. Direct observation of facts, unfortunately, may become secondary.

The awareness of the contrasts between illusion and reality set me onto a path of writing a book I initially called *The Social Illusion.* My original idea for the book was somewhat generic. Gradually, in the process of researching and writing for more than a decade, all sorts of things changed. My attention broadened to include not just the interpersonal and societal but also the global. The overriding role of hierarchy, communication, and political economy in society became more apparent. Most importantly, I became increasingly aware of the complex role that both fear and habit play. Either one can trigger illusion or play on our failure to accept the overwhelming evidence of the converging global crises that confront humanity today.

I grew up in an environment that encouraged me to explore the world and to value discovery. My university training in philosophy of science, statistics, and research methodology added to that outlook. It taught me to build factual knowledge based on observation, careful measurement, and systematic analysis—and to try always to get a feeling for the reality the data represents. That requires being willing to give up beliefs that do not pan out when subjected to the rigors of scientific observation and analysis. To purge illusions from our understanding of reality became my quest. I came to value verifiable fact over any belief, including my own, although I understood that the distinction is not always so easy to make.

When I began that quest, I could not have guessed where it would eventually take me. I met many smart people along the way, each of whose understanding of the modern world was radically different from that of some others. It is not that the world is unknowable or that only the smartest people know the truth. Rather, the perspective each of us forms from our own experiences never completely separates illusion from reality.

Nobody is exempt from the illusions that form part of our understanding of reality. That is why we are sometimes forced to change our minds. However, most of us hold tightly to beliefs that have reassured us for a long time or simply worked well. We develop such beliefs within the groups with which we most closely identify.

Nevertheless, a changing reality has a nasty habit of ignoring our images of it, no matter how long-standing they may be. If we live entirely within the language of our group, we may never notice its inconsistencies due to realities we may not have personally experienced. When confronted with changing realities, and there are many, it is up to us to either get real or suffer the consequences.

Of course, for too many people, reality is an entirely settled matter, even though they may not actually understand much, if anything, about it. Something has limited their range of wonder. Some may know a lot, especially in some technical fields, but understand very little due to a lack of wonder and a narrow scope of vision. Such folks tend to hold onto an absolute belief system no matter what, which is quite comforting in its simplicity, certainty, and finality—if not accuracy.

Factual challenges to absolute belief systems can be quite disturbing. To the extent that they acknowledge them, the changing facts of modern existence enrage many adherents to rigid belief systems, whether in the US, the Middle East, or elsewhere. In present-day Iran and among adherents to some fundamentalist groups in the US and elsewhere, members simply do not tolerate such factual challenges. Unfortunately, for that reason, "true believers" are much more susceptible to political manipulation than are folks with wondering, skeptical minds. We definitely need more wonder to offset certainty grounded in nothing but itself.

The history of demagoguery reveals its consistent exploitation of rigid belief systems. In that context, I understand that political belief systems—both ideologies and utopias—rest too much on the fear-suppressing illusions of certainty—and not enough on facts. Unfortunately, ideological certainty often spawns a belligerent refusal to change beliefs when confronted with facts. Certainty is comforting; often reality is not.

The clash of beliefs drives a lot of conflict in the world. Yet, behind most ideological conflicts, we often find a power struggle. Order and conflict may be the two concepts most central to sociology. Understanding human affairs always requires that we look to the underlying relations between the dominant power and its attached belief system. Who controls the social order, and how do they maintain power? Who controls the flow of information and to what extent? Who challenges the social order, and how does conflict evolve? Those questions seem central to how the world works or how it does not. Maybe that is what drew me to the social sciences in the first place. Knowledge seems more tentative than say, basic if not quantum, physics. Certainty remains always tentative.

My research for this book began with the desire to understand societal belief systems and their relations to whatever controls the outcomes of everyday life and the machinations of political economies. It continued as I explored the catastrophic convergence of crises in both Nature and society that confronts us today. That naturally led me to questions about how humans would or would not, respond, or do not, to the globalized New Great Transformation these crises forced onto the world.

Most of today's crises, if not all, are consciously or unconsciously of our own making. We think we know how the world works even as it changes before our eyes. Stable knowledge reflects stable conditions until those conditions change. Change reflects conflict, either between groups or between a group and the conditions under which it lives. Sometimes, such as today, we are the unwitting (or witting) agents of the changes that we fear. Knowledge built up over centuries may no longer apply, especially when the new conditions under which we live conflict with the lifeways and knowledge that we have embraced for a long time.

Today, humanity faces challenges unprecedented in history. We are living through the end of the Industrial Age—some say that will cause the collapse of civilization itself. In any case, the current convergence of catastrophic changes in the world is rapidly becoming global in scale, and the crises we face are mostly human caused. Yet, humanity is ill-prepared to confront them. "Leaders" insist on treating these catastrophic changes as mere "problems" within the outmoded model of "business-as-usual."

The only way we can come out of this alive is to understand what controls the changes we have caused and to respond to them quickly, decisively, and comprehensively. That means abandoning the economic and social illusions that got us into this mess in the first place—a daunting task at best. That is why we must understand how we know things and how that affects what we do.

What We Do and Why

What were you doing yesterday afternoon? Shopping after work for groceries or a new electronic toy, playing a pickup game of hoops or doing your homework, meeting with your church group, having a pint at the pub after work, paying your mechanic the high price for repairing your car, or just watching TV or playing video games? Or were you plotting the assassination of a political enemy?

Well, if you were doing that, you probably would never have heard of this book, although most folks would not have heard of it even if it had achieved bestseller status. Why? Because most human activity in modern industrial civilization focuses on the various consumer products and activities that characterize the consumer culture. Most of those who do read books have a novel, a cookbook, or some kind of self-help book in their hands. Nonfiction books discussing major issues are not nearly as popular unless written by a political celebrity or media star. And why is that?

If you look at the above activities, which I just wrote down as they came to me by visualizing what folks might do after work, you will see that almost every one of them involves some industrial-consumer product or service. That is what participating in an industrial-consumer culture is all about. However, we American "individualists" perceive most of our actions as expressing some element of our personal identity, not as an element of the complex system in which it occurs. Actually, we perceive our actions as individual or personal because most of us have unconsciously taken on some elements of consumer culture as part of our identity. Forget that myth of the "rugged individualist," which is a cover story for the collective conformity of the consumer culture.

Behavior aligns closely with identity. So, we shop, we consume, and we throw away a lot of stuff, trying to "express ourselves." I remember, even back in high school, a lot of guys put various stickers onto the windows of their cars advertising the custom parts they used in their hot-rods. Some put on the emblems for their favorite pro football team or an image of their high school mascot. People identify with the products or services (especially the service of getting to watch their favorite sports team) in which they indulge themselves. Most of us identify with the place we live, whether in "the hood," a suburb, or a plush, gated community. We also identify with our family, of course, our state versus others, and especially with our nation.

Behavior reflects personal identity, but it also expresses our relations with other persons, groups, and institutions. Most of us are participants in each of these levels of social organization in some way; thus, we act in various forms of coordination with others. That is, we cooperate with others in a complex web of relations that make up the social order. Since this web of social relations tends to last over time, our patterns of action tend to be habitual. We develop habits, and habits are hard to break. In fact, some habits are so strong that we might call them behavioral addictions. Beliefs can become habitual in much the same way as actions. They develop our perception of our significant others and become as strong as those relationships. Some beliefs can become addictions too.

In *The Power of Habit* (2012), Charles Duhigg makes a strong case based on a great deal of research that habit and beliefs drive much of what we do, though we believe we are making rational decisions with each action. Decades of social psychological research demonstrate how habits of both action and belief arise, as well as why changing either or both is so difficult. We find little time, if any, to reflect upon where we think we are going. Yet, our quest continues as if its goal were clear and certain.

Now, this may sound too negative, as if habit or automation controls all human behavior, but it is not. We must remember that behavior aligns closely with perception and memory too. It is a far more complicated matter than it would seem on the surface, as a great part of our behavior involves what we did in the past and whether or not it worked. Tradition, formal or informal, is also a kind of habit; it plays a big part in what we do. But independent, critical thinking can be violated by habit too.

Habits have a lot to do with stability of behavior, whether good or bad. Of course, some bad habits can result in a downward spiral into instability and ruin, such as compulsive gambling, alcoholism, or drug addiction. Other habits, which are usually the internalization of widely accepted social norms, result in personal and social stability. That leads to the ability to predict behavior about the consequential matters in life. We thrive in society to the extent that we can predict a lot of the behavior of others.

One of the two things that stuck with me most from my first college course in social psychology was that the reasons we do what we do we often formulate after the doing. We often just act and then we explain to ourselves and others why we acted as we did. It often happens so fast that we don't notice the sequence. That is not so surprising when you consider that experience, tradition, and habit play a big role in determining actions. In a particular situation, so may the pressure we feel from friends or even enemies. The expectations of others always form part of the background to our actions. Often, we simply know how we will act in a given situation even without thinking about it. That is why it is usually easy to explain our behavior even though we had not thought it out ahead.

It also stuck with me from that first social psychology class that if you want to persuade someone of something, you had better speak their language. No, I'm not just talking about English or Spanish or Vietnamese; I am talking about the language of motivation and belief. Everyone has their own narrative (story) about not only who they are but also about why they do one thing and not another. It explains what we do to others.

To communicate very effectively with another person, you must understand their belief system and what values, hopes, and fears lie behind it. To understand who they hate, love, or respect and why is very important too. These factors lie behind and strongly influence beliefs and behavior, whether or not the person is thinking about them at that moment when they act or refuse to act when called upon by others. How you expect another to act, or not, can be at least as important as the act itself or the rationale for the action presented. Whom do you trust?

One of the benefits of the physical sciences for social scientists is that they show us how our behavior and interactions work beneath the social level. For example, the neurosciences have come a long way in recent years, due to advanced computing techniques, as well as sensing and measurement systems. (They can measure and visually trace minute processes in the human brain.)

The processes of perception and cognition are especially fascinating. We now know with certainty that the brain filters and processes vast quantities of bits of data to form information that makes sense to us. We unconsciously organize our conscious mind around if and how a bit of data—now a signal—fits with whatever we already know and how we know it. All our knowledge and prior experience, to which that bit of information may relate in some way, play a part in if or how our brain ultimately interprets or ignores it.

The idea that "That's unbelievable!" can actually play out at the neurological level even before reaching our consciousness. The brain filters vast amounts of raw sensory data to glean any meaning out of our ongoing flow of everyday perceptions. It ignores far more data than it incorporates in shaping our awareness and the meaning of our experiences. In doing so, it relies on its experience of past patterns of data and their meaning.

These neurological processes are far more complex than any supercomputer or network of computers. Yet, networked computing, automated financial systems, and most recently artificial intelligence appear to be overtaking human decision-making in many sectors of life. This trend is evident in the globalized corporate political economy of growth that drives almost every nation today.

Nørretranders (1999) wrote a fascinating book on the neuroscience of meaning, going into a great deal of detail to explain why our simple conceptions of ourselves, and our computers as well, do not reflect the richness of our lived experience. He argues that the modern information system, flowing through the many networks we call the Internet, is now "an autonomous organization … proliferating, with its own logic beyond our intentions." (p. 362) But that is another fascinating story. The most important takeaway that Nørretranders provides is the value of understanding human consciousness.

Human consciousness is capable of amazing things. Yet, the modern organizational penchant to plan and regulate everything in the interests of efficient economic growth and capital accumulation runs counter to everything we have learned about our complex connections with the ecological systems that sustain us. This approach has set industrial civilization on a path of habitat and self-destruction—risking our lives. It would appear that our beliefs about what we do in our industrial-consumer culture and its effects on the world are both very mistaken, despite the temporary robust results of material production. How could our collective consciousness have gone so wrong?

Part of the problem is the conflict between the complexity of human nature and the simplicity of the behavior that the institutions of the Industrial Age typically call upon us to do. In the drive for organizational efficiency, all sorts of complex institutions require simple individual behavior. But humans are not simple; we are very complex creatures. John M. Gowdy (2021) wrote a fascinating book,

Ultrasocial, in which he describes this conflict and its consequences for the future. He argues that the global political economy is a super-organism that requires human subordination in ways that sacrifice individual—and I would add community—wellbeing for the benefit of the global market.

Social control is a complex process that can happen in many alternative ways and lead to a plethora of outcomes from wonderful to disastrous. All we need to do is look at historical instances of societies that flourished for a while and then collapsed because of their own misdirected intentions.

Social control results from the cooperation of persons and groups within society, sometimes under the duress of a tyrannical regime. Other times, it occurs under an umbrella of consensus and in incredibly multifaceted ways.

Today, we desperately need to create a consensual form of rational societal transformation for our failing global-corporate-political economy. This transformation is crucial so that humans can live in harmony within what remains of the habitat we have already nearly destroyed. To create that, we must understand mutual aid, cooperation, networking, integrated social control in social groups, and society within the context of Earth-System limits. These necessities directly conflict with the superorganism's demands for simple behavioral compliance with its operational demands.

Fly Boy and the Culture Clash

My high school history teacher was a retired commander in the Navy. Just before graduation, he convinced me to join the Naval Air Reserve, which sounded good because of my lifelong infatuation with aviation. The military draft was active, but no major military conflict really threatened my future choices at the time. Vietnam was not even a glimmer in Robert McNamara's eye. I signed up. It sounded like an adventure, and, of course, it was. I went through boot camp at the Los Alamitos Naval Air Station south of Long Beach, California. The discipline of the Boy Scouts was nothing compared to boot camp. Yet, Basic Training was not a terrible challenge either: just learn what you are supposed to do, how to do it, and how to follow orders. I learned a lot about the military culture and about all sorts of technology too.

However, the only flying I had an opportunity to do was hitching a ride with pilots who had to get in their hours in the air for flying pay. We usually flew in a Gooney Bird, the C-47 military version of the civilian DC-3 transport, one of the most reliable and venerable aircraft ever built. I experienced a variety of work assignments, some extremely noisy and dangerous, such as on the flight deck during a two-week training cruise on an aircraft carrier. The USS *Kearsarge* was an Essex-class aircraft carrier built right after the end of World War II. They gave us trainees a chance to fire a M1918 Browning automatic rifle off the stern of the ship, aiming at the garbage that the galley crew routinely disposed of in that way. Even then, that small act of pollution disturbed me. I knew somehow that collectively that kind of "waste disposal" writ large could not end well. Today single-use plastic strangles sea life.

But I really wanted to fly, as in piloting airplanes, not sitting in the back observing the action. I checked out the flying opportunities in the Air Force, Army, and Navy, then applied for officer/pilot training in the Air Force. The timing was not right, or maybe it was in terms of my survival.

This was happening a few years before the US got heavily involved in the Vietnam War. The Air Force had determined that they had too many pilots. My acceptance into the program meant nothing because they stopped training pilots with the class I could have entered if I had not delayed entry to finish the semester at El Camino College. Later, I discovered that every cadet in the potential class I could have joined, even the top student pilots, had been washed out because they were deemed unnecessary in the pilot and aircraft inventory. Had the Air Force decision-makers realized what was coming—the Vietnam conflict—I might have ended up in the Hanoi Hilton with John McCain.

Disappointed but desperate to get out of my parents' house, I signed up for officer-navigator training in the Air Force. After a few months, I realized that I did not want to spend the next four years sitting in the back seat of whatever aircraft someone might assign me to—or to be the bombardier targeting a nuclear device over Moscow. So, I resigned and served the obligatory two years enlisted as a geodetic surveyor, measuring the precise location of Minuteman ballistic missile sites in Montana—you can't hit your target if you don't know precisely where you are firing from. It seemed I could not escape the Nuclear Age. Nevertheless, I learned a lot about military culture, its hierarchical organization, and the skillset of geodetic surveying. That real-world application of trigonometry to surveying and artillery is technically much the same as navigation.

The socialization of military training tends to stick with you. A couple of years later, my roommate at the University of California, Santa Barbara, pointed out certain behaviors, such as how I arranged my socks in the dresser drawer, that were unconscious hangovers from the rigid requirements for just about every behavior in officer training. I realized then that my military experience, limited as it was, set me apart in some ways from my fellow students, most of whom had gone directly from home and high school to their college dorms. What does that tell you about habits and social control?

Coordination and Integration in Social Groups

Several years after my military officer training experience, I completed a PhD in sociology. It allowed me to continue seeking answers to the mystery of how human systems can hold it together—and why sometimes they do not. Once settled into my university career, I finally achieved my childhood dream of flying by taking lessons at an airport near the campus where I taught.

When I began flight training, I immediately sensed the importance of self-control to execute the precise procedures required to effectively operate any aircraft and stay alive. It is so much more complicated than driving a car. A couple of years later, in the process of earning the instrument rating that authorizes one to fly on an instrument flight plan and make approaches to landing in clouds, using only instruments to navigate, I experienced a completely new level of precision, discipline, and skill.

There is no room for error on the final approach to minimums at an airport shrouded in clouds and turbulence. Shortly after attaining my IFR rating, I flew my first instrument approach to minimums after flying a holding pattern with altitude step-downs at several levels for over a half hour, all the time in the clouds. Talk about having to concentrate! Once, flying into the Lake Havasu airport for fuel on the way from Santa Fe to Los Angeles, as I began to descend for landing, I experienced extreme pain in my sinus, which had clogged up from allergies. When landing an airplane, no amount of pain is relevant; I simply had to suppress my awareness of the pain on that very hot day in the desert.

Several years later, my wife graciously gave me five hours of helicopter training as a birthday present. I experienced developing a skill set several orders of magnitude more complex than flying a fixed-wing aircraft. A few years later, as another birthday present, I got a feel for the experience of air combat through a mock dogfight in a Sai-Marchetti Italian fighter trainer, with an instructor in the cockpit with me. I loved the aerobatics and was happy that I shot down my competitor, simulated with lasers, more times than he shot me down.

Soon after that, watching a documentary about the Air Force Thunderbird aerobatic team fascinated me, of course. In some ways, jet fighters are not as complex to fly as helicopters, except perhaps when landing on an aircraft carrier in turbulent weather and high seas. Alas, aerobatics and helicopters were fun adventures but out of my price range for a regular diet. These are skills maintained only by constant practice.

The precision flying of an aerobatic team is quite another level of human performance in coordination with others. That documentary not only showed the epitome of the precise skill development in each pilot but highly refined coordination with the others as well. It also revealed the social implications of the workings of the small, dedicated Thunderbird organization, as well as the personal dedication and extreme coordination required of its members.

High-speed synchronized team aerobatics is an activity that requires absolute and precise social control through the interpersonal coordination of every member of the team. Extreme coordination of personal self-control in the air at high speeds becomes precise social control. Technical precision and behavioral coordination are mission critical. Each pilot must behave in entirely predicted ways, as wing tips may be just a few feet apart in both practice flights and at public air shows. A slight deviation can mean death and, indeed, sometimes has.

In this context, individualism is both unacceptable and flat-out dangerous. It must always be left on the ground—a feat in itself for such high achievers. Mutual trust is paramount. Strict compliance with the agreed sequence of operations and precise timing of their execution becomes a highly personal commitment for each team member. This includes a high level of physical training and disciplined bodily control. Such training enables pilots to avoid passing out under extreme G-forces that would otherwise drain blood from the brain while, for example, pulling out of a steep dive or turn.

It is interesting to note that such surrender to the requirements of group action also involves a great deal of individual self-esteem. The performance of group participation involves a level of commitment that reflects very strong egos. While the high level of top gun performance leaves no room for individual deviation from the norms of aerobatic procedure, the requirements of the job also necessitate a very strong character, which some express quite energetically in other ways on the ground.

Typically, members of social groups internalize social control, some more so than others. For example, each of the pilots in the Air Force Thunderbirds and the Navy Blue Angels fully internalize a very strict form of social control. They exercise total self-control in the exact terms of the group's requirements, which, in a way, is also an expression of their very strong egos.

The highly precise and conforming performance of the individual validates the ego in the context of the metaphorical superego, which represents the moral-behavioral standards of the group and its mission. In that, we can see some parallel in the actions of a street gang in south Los Angeles, although my knowledge of that milieu is entirely secondhand.

The operations of specialized groups such as the Thunderbirds epitomize the highest form of military discipline realized in each pilot's individual performance in precise coordination with teammates. The pilots execute idealized versions of air combat skills that, however, never quite reflect the actual messiness of real-world military operations and combat. Nevertheless, the achievement of exercising a precise maneuver provides the control skills necessary to respond quickly and accurately to the complexities of actual combat. That is what training is all about.

When the battle begins, every nanosecond a new situation arises in which pilots must temper the skills of flight with creativity. They must telescope strategy into the tactical instant. That is why no contender ever met Colonel John (Forty-Second) Boyd's challenge to other jet fighter pilots to shoot him down in mock combat. None could. Boyd was both a master strategist and a technically precise tactician able to maneuver his aircraft more effectively than any challenger could. He viewed combat as a complex dynamic system of "Destruction and Creation," the title of a brief unpublished paper he wrote, which eventually became the basis for the US Marine Corps doctrine of maneuver warfare. Boyd understood the complexities of system dynamics on the fly. Most of us, including police, military, and other first responders, can only aspire to exercise such an extreme level of complex competence.

Unlike the linear thinking behind traditional military and police theories, Boyd's model recognized the nonlinear processes involved in combat dynamics (Coram 2002). Yet, the dynamics of creation and destruction apply to *all* complex adaptive systems, from interactive episodes such as aerial combat, to events such as business meetings or police-citizen contacts, to the functioning of entire societies. The primary difference is in the speed at which change occurs.

An evolving techno-industrial economy recreates itself in part by destroying existing economic formations in the process of building new ones. (Schumpeter [1942] 1975) As humanity moves further into the Anthropocene, conditions of all kinds change far more rapidly than in the past. This calls for equally rapid personal and societal changes involving responses of destruction and creation we can barely imagine but must prepare to carry out very soon—as in now.

The key factor, just as in aerial combat, is the ability to adapt and overcome new and unexpected factors that may determine ultimate survival, whether of the individual, the group—or even the species. In combating global heating and its consequences, we get no practice runs, no do-overs. Our real-world responses are our training, and we must rely on the ability to recognize and even predict change and act within that complex dynamic. Here is where our habits can become our enemies.

Some circumstances call upon us to participate in very strict ways as members of social groups, formal organizations, or just as citizens of the larger society. We comply willingly or resentfully, we rebel impulsively, or we revolt intentionally. In officer training, I had complied with military commands effectively, simply by exercising personal self-discipline. The parameters were quite clear, so why buck a system you joined willingly but cannot control? Was that stoicism? Perhaps. Effective performance and survival require us to recognize and distinguish between parameters we cannot control and elements of the situation that may allow us to achieve a goal, sometimes by systemic process intervention. In survival situations, that always means being open to new interpretations of reality, for reality has changed, and habits can become dangerous.

I have always questioned the social arrangements by which people live, although I did not always know why. Today, we know that the relatively stable conditions we became so accustomed to during most of the eleven thousand years or so of human development during the Holocene Epoch are changing rapidly. That rapid change is accelerating, though not in a good way. Hierarchical, linear social systems worked relatively well during much of the Holocene, especially for elites, as political economies expanded into new relatively pristine territories—sometimes producing empires—long before approaching planetary or local-habitat limits. Economic expansion also gave individuals a false sense of personal freedom in the context of new opportunities in a hierarchical system that promoted ideas of liberty and expansion. It is easy to dominate when you have superior technology and many resources in a field open to expansion. That situation is over.

However, in the context of changing conditions, distributed decision-making within systems is more effective, as demonstrated by Boyd's theory of maneuver warfare. As any experienced warrior knows, as soon as the shooting starts, all linear battle plans evaporate. Combatants must respond to the immediacy of rapidly changing conditions all around them.

The military structure is a highly stylized formal expression of hierarchy. It applies discipline to maximize control of its members to effectively implement its strategy through centrally controlled tactical precision. The goal is to sustain that control under the highly stressful and dangerous conditions of combat. Yet, combat presents a complex dynamic that no military hierarchy can ever fully control. When the battle begins, all bets are off and the nonlinear (nonhierarchical) principles of responding to multiple complex system dynamics prevail. That is why Boyd's multidimensional combat skills prevailed over all challengers. The same dilemma of the need for a nonlinear response exists for all of us in the hyper-dynamic world today, which the military, political, and industrial hierarchies usually define as linear.

The dominant culture of the industrial-consumer global economy is hierarchical and linear. However, the modern world presents itself, in all its complexity, to the current hierarchy of the globalized political economy in ways that are simply out of reach for linear thinkers. The formalisms of the international "rules-based world order," and domestic law enforcement as well, fail under conditions of accelerating change.

In officer training, I had not fully internalized the culture of military discipline as much as I complied with it from a practical perspective. I performed according to its requirements, but it had not fully taken hold of my consciousness, or so I thought. Anyone who has performed well in training has internalized some level of those behavioral forms.

Years later, I would discover reflections of that formal military discipline in certain behaviors I had unconsciously retained as habits. Internalized social control is most powerful and effective when skills become habitually deployable. Yet, the resulting capacity for self-control need not necessarily restrict creativity. Boyd's creativity (his nonlinear response to complex rapidly changing conditions) made full use of his trained capabilities in order to exercise a very creative kind of dynamic self-control. Hmmm …

In a similar way, martial arts are both formal and dynamic. In practicing Aikido, by the time I reached the rank of Shodan (Black Belt), I had developed ingrained patterns of response to potential attacks. However, to be effective those responses require precise situational awareness, that is, a response to an attack must be attuned to the exact motions of the attacker. Otherwise, the move will not work. When it does work, it feels as if nothing happened. That is because the response perfectly blended with the energy of the attack. Oddly, it might seem that if humanity is to dig ourselves out of this downward spiral toward societal collapse, we must respond ever so precisely *and* creatively to the attack in relation to climate/ecological destabilization. We have no time to practice, and our response must be exquisitely accurate.

Processes of creation and destruction occur all the time within social groups and complex organizations and throughout Nature. They can accelerate in the context of changing conditions, although hierarchical restrictions in formal organizations—bureaucracies—may constrain them. Situational dynamics always clash with hierarchy. Some social groups buzz along smoothly. Others fall apart from failures of commitment, denial, indifference to changing conditions, or a lack of discipline, coordination, resources, creativity, or vision. Yet, in the US at least, it is usually framed in terms of individualism.

In the US, we value what we see as our personal freedom, sometimes to the extreme. At the same time, we expect the security that a stable social order provides. We often expect others to conform to a certain image of correctness but do not always hold ourselves to quite the same standards. When a community, or any social unit, grounds itself in mutual values and a high level of common purpose and consensus, things tend to go smoothly, even when confronting a crisis, especially when collective commitment is high.

Today, we face the challenge of exercising creative destruction to rapidly reduce carbon emissions and ecological destruction, and to transform economies to stop destroying our habitat. To meet that challenge, we must achieve a high level of consensus on values and goals to create a livable place for us in the early Anthropocene—no small task. On that front, we have a long way to go in a very short time.

So far, we have no broad societal consensus for reaching such a goal, either within or among nations. That is also the case with several other major social issues. In this context, political platitudes do not count. The role of law enforcement and the military under the changing societal and ecological conditions of this emerging New Great Transformation of our global and local habitats will be problematic at best. Genuine social order does not emanate primarily from law enforcement; it results from sufficient societal consensus to sustain it. However, that societal consensus is increasingly in very short supply precisely when we need it the most. That is one very big reason why law enforcement has such a difficult task.

We live in faltering complex social systems, which, as societies, we have not held together well. Now we must transform society in unprecedented ways that we do not yet fully understand. In Chapter 3, we will examine the nature of complex adaptive systems, in particular the human kind. We must understand the fundamentals of how to achieve our aspirations for a human future by transforming our social institutions.

Yet, in all of this discussion, something is still missing. Human evolution worked over millennia because of cooperation and mutual aid. In other words, community is a core human characteristic. Here is where a relatively new concept—for what was once a far more common practice among human groups—can help in a world unlike anything our ancestors ever experienced. John Brown Childs (2023) uses the term transcommunality to describe a method for overcoming the fragmentation of modern humanity and coalescing diverse peoples into a viable alliance for change.

Childs borrows the concept (if not the word) from the Haudenosaunee, or Iroquois federation, in order to frame our thinking about communities as coalitions rather than as collections of like individuals in a simple group. Thus, he takes into account the differences in the very act of coming together. He describes the Iroquois in terms of transcommunality in the actions and the outcomes that many people desire in seeking social change, such as racial justice or peace. The focus is on how many disparate groups may organize without questions of group identity interfering with the goals of organizing a coalition or alliance.

There is no simple map to success in this complex process of achieving higher goals in the context of diverse groups. No pure ideology will do the job. That would lead to dogmatism and often exclusion of those who do not meet impossible standards of compliance or subordination. The underlying values are trust and respect— values that are in rather short supply in the culture of industrial modernism, where competing identities of people whose traditional social anchors no longer exist.

In the examples I gave above, high-performance groups rely on specialized forms of consensus and mutual commitment to a goal within the context of a particular hierarchic organization to achieve precise coordination. In contrast, diverse social groups seeking to resist the oppression of globalized monoculture—to attain freedom from the modern hierarchies of domination—must find another way to achieve effective concerted action. The values of transcommunality offer that opportunity. Childs offers two simple formulations to illustrate the contrast between the oppressive monoculture of the modern global political economy and the potential for people to resist and achieve the societal changes necessary to create an ecological civilization.

HOMOGINAIZATION + FRAGMENTATION =
SUBORDINATION

This simple formula represents the culture of domination that has ruled society for hundreds of years, even longer. On the other hand, an inclusive model respects diversity, allowing it to work.

HETEROGENEITY + COOPERATION =
RESISTANCE AND FREEDOM

Transcommunal cooperation transcends lines of social identity, such as race, class, or gender, by recognizing a higher value in humanizing "the other." Creativity blossoms under such conditions, enriching coordination in and among groups seeking wellbeing for all. It is interesting to note a parallel here with evolutionary biology. Diversity is a key to the enhancement of life through biological evolution (DeFries 2021). It is also necessary for generating the creativity needed for enacting social change.

Survival of the fittest humans often means the fittest social group, which does not necessarily imply competition with another group as much as between members. More important is the fitness of the group's culture to work within the habitat in which the group lives. Today, it is most urgent to meet the challenges of the Earth-System changes we have caused. That requires us to facilitate a new form of cooperation to achieve societal stability within a rapidly changing bio-system context. Creativity is the key.

CHAPTER THREE

Systems and Structure

DURING MY FIRST year in graduate school, I was casting about for a topic for my master's thesis. By then, I was heavily into social psychology. I studied it as an undergraduate with Tomatsu Shibutani at the University of California at Santa Barbara. Human behavior was beginning to make sense, at least in some ways.

Shibutani was an amazing teacher; his classes were always crowded. He walked into the lecture hall with his hands in his pockets, went to the blackboard, wrote out an outline, turned to the large audience of students, and began to talk. He rarely looked back at the blackboard as he paced around and explained its content in detail. If you took good notes, they would become a complete elaboration of the outline he had written on the board. Obviously, his impeccably organized mind expressed in his lectures much of the information in the social psychology textbook he had published. He provided his students with a bibliography that was several pages long and expected us to read as many of the works it cited as we could. He expected us to choose whichever works drew our interest the most and use them as the basis for our term papers. He also expected us to exercise intellectual judgment, not just regurgitate memorized facts or theories.

With the Shibutani experience as context, I searched for a thesis topic. The relationship between individuals interacting and the social structure in which they participated interested me. One of my professors offered me a connection that allowed me to observe the inner workings of a small urban bureaucracy in Saint Louis, Missouri. I spent the summer there observing behavior that I did not yet understand.

Meanwhile, I stumbled across some literature reporting early attempts to formalize social network analysis. This required applying directed graph theory from physics and using matrix algebra to analyze multiple patterns of interaction among members of a network to produce graphs depicting the configurations of their relations. In other words, I documented and analyzed the structure of the network. The idea was to observe interactions between members of the group and note the character of their relations. Then, I graphed them and their characteristics in the form of a network of nodes and connecting lines based on the data analysis. I looked for some meaning in all that, such as the degree of centrality of a node (person), how well connected they were to other nodes, social distance between nodes, density of relations, clustering, etc. It was fascinating, though difficult to accomplish.

None of my professors knew anything about this stuff, which was both a blessing and a curse for me. In judging my work, the members of my thesis committee would have a hard time criticizing technical analysis they did not understand, but I had to explain it to them. I also had to show them that it was valid and could be useful as a tool in social psychology. At the time, its potential had some technical limitations.

That period, around 1966–1967, was the early days of computing when the big early mainframes were far less powerful than laptops are today. For my analysis of a small network, I needed a mainframe programmer at the university's computer center to write code applying matrix algebra to the data based on the interaction concepts I was investigating. I coded the data on IBM punch cards—does anyone remember those? I sent a tray of program and data cards to the computer center, and a week or so later, the printout came back with printed garbage because one IBM card was out of place. It took a couple of runs to get accurate, meaningful output. Those large mainframe computers barely had enough power to do the job.

Anyway, the project finally worked out, and my thesis committee approved it. But, for me, the exercise demonstrated that the complexities of social structure in a small group can be huge, but they can be analyzed objectively only if you have sufficient computing power. Back then, I could use most of the processing power of the mainframe campus computers for a fairly simple analysis of a very small social network.

Today, it is a completely different story. Desktop computers can handle much larger data sets to analyze social structure, and the field of social network analysis is becoming very robust with the help of vastly more powerful computers and software. Off-the-shelf software is available for doing social network analysis. The structure of social networks is gaining importance in sociology due to increased computing power for sophisticated analysis of social group organization and control mechanisms within and between networks, depending on interaction structures.

What is the point, you ask? Well, I think that social network science will soon provide insights for further understanding complex adaptive human systems, how they evolve, and, most importantly, how they can intentionally change. That is the point: we—modern societies—must change in ways we have not yet figured out. Anthropologists, archeologists, and evolutionary biologists today understand so much more about the structure of early human groups and how our ancestors survived by cooperating and evolving cultures that worked effectively within their habitats. All that happened within relatively small social networks over millennia.

Cultural means for cooperating allowed groups to exercise social control in ways that assured their survival. They were not just individual "ape-men" individually running around grunting and throwing spears at mastodons. Small bands of hunter-gatherers worked together to obtain enough food for survival, using simple tools in sophisticated ways. They were able to spend much more leisure time enjoying each other's company and telling stories as they ate the day's bounty than modern corporate workers have time to microwave a TV dinner today. That alone is certainly food for thought.

None of this is to say that pre-modern peoples had it easy or even had very long lifespans. Life "in the wild" was tough. We moderns do not have the tools—technical, psychological, or social—to survive and prosper in a similar way. Look at what our allegedly sophisticated technology and bureaucratic organization have done to the natural world. We mostly ignore it as something outside and separate, which we have only considered for extracting materials and fuel, as well as providing a dump for our waste—until now. The habitat we are destroying provided a living for our ancestors.

Societies hold it together in different ways under various conditions, unless they do not. When you get right down to the basics, certain principles apply. One principle is the need for some degree of functional differentiation. That simply means that some skilled members hold responsibility for specific functions and others handle different functions. Industrial societies have carried that principle of organization to the extreme. Coordination between operational functions is also important but is more difficult as the complexity of the organization grows. I could not count the organizations I have seen fail because of poor coordination amid cross-purposes—especially when some egos do not play well with others.

At the same time, the group must maintain a significant level of social cohesion for it all to work. In eking out a living directly from a harsh environment, every group needs a good deal of cooperation and a deep knowledge of its habitat. How is that possible? Some argue that competition for food and other survival requirements is what drives human behavior. Such speculation overgeneralizes from just one element in Darwin's theory of evolution. Survival of the fittest humans often means the fittest social group, which does not necessarily imply competition with another group as much as fitness of the group's culture to work within the habitat in which the group lives. That most often results from cooperation within the group.

Evolutionary biologists have come a long way in extending and refining elements of Darwin's theory. They use extensive evidence from modern studies in paleontology and archeology, often using diverse advanced technologies, such as Lidar ground penetrating radar and DNA analysis. While competition exists and can be determinative in some situations, the core feature of human social organization that allowed us to survive and prosper all these thousands of years is the cooperation among members of locally situated human groups. Humans are one of the most social animals on the planet, and our social relations may be our most powerful survival strategy—if we recognize and use that tool effectively.

We can demonstrate this by the actual lives of hunter-gatherer bands. Few such groups exist anymore. However, in the 1950s, some Harvard anthropologists recorded on film in great detail the lives of a band of Bushmen of the Kalahari Desert in southern Africa—before the forces of industrial modernity (and the diamond trade) changed their lives forever.

Watching that classic ethnographic film in an undergraduate anthropology class many years ago made a lasting impression on me. I remember it vividly (in black and white) even today, decades later. The bushmen of the Kalahari Desert in southern Africa were called the !Kung or the San. In the harsh savanna environment where hunting and gathering were the only means of provisioning themselves with the necessities of survival, they lived in small bands of a dozen or so. The film showed the details of the lives of a small band of !Kung bushmen.

What first caught my attention about the Bushmen in that film was that it depicted how they lived before the intrusion of modernity. The filmmakers shot the film in 1952–1953, when the lives of the !Kung people were still relatively undisturbed by outside forces, partially because diamonds had not yet been discovered in their habitat. These folks lived a hard life, certainly by our consumerist standards, hunting and gathering their food day by day. Yet, their cheerfulness and calm were striking. The !Kung people lived in a world that they fully understood and accepted. Our so-called "lifestyle choices" were simply not an issue.

The lifeways of a group living in such close contact with each other and the immediate ecosystem on which they depended for survival can be more illuminating for us "moderns" than we might expect. The social organization and interpersonal relations of the !Kung harmonized eloquently with the environment that sustained them; they had to. And it worked internally as well.

Each had a role that was important for the group, and the group sustained each individual. They survived on mutual aid, skills, and cooperation. Their social organization met in intricate ways the requirements and exigencies of surviving in the thirty-five-thousand square-mile, semiarid savanna they inhabited. They knew and worked with the key resources available to them. They did not, nor could they, change the basic parameters of the environment of which they were an intimate part.

That world has since changed radically. Yet, life in the Kalahari in the early 1950s remained as it was for eons before incursions by Europeans who brought the forces of extractive capital—they found diamonds. Their experience may have important lessons for societies facing the emerging New Great Transformation of climate, eco-systems, and our industrial-consumer society itself. The !Kung no longer live as they had for so long; European invaders forced them from their habitat. Now, most are agrarians and pastoralists; some occasionally work for cash.

Neither can we industrial people live as we have for the last two hundred years or so. Reflecting on the vast differences between the lives of the San people and our own made me question our easy assumptions about what is natural and what is right. Our cultural detachment from Nature raises the question of what progress among humans might actually be in the context of the faltering industrial-consumer economy in the increasingly unstable conditions of the early twenty-first century. The title of the film that fascinated me was *The Hunters* (Marshal 1957). A good part of the film followed a group of four men tracking a giraffe over a period of several days with nothing but spears as weapons.

Not only were their skills and persistence amazing but also their cooperative interactions revealed certain very important complemen-tary opposite traits of the men as they interacted among themselves. They carefully coordinated their actions on the hunt. The group consisted of a simple network of individuals enacting the roles of the headman or chief, the shaman, the warrior (technical expert), and the clown—yes, the clown. Their roles were primarily functional, reciprocal, and interpersonal, not hierarchical. Through their coordinated action, after several days of tracking, they finally killed the giraffe. That kill became a boon to the survival of their little band and that of their neighboring bands for weeks.

On their return to camp, the hunters shared widely the rewards of their successful hunt, truly a group effort, which had brought them nearly to the point of exhaustion. Great celebrations of the successful hunt went on for days. The kill provided far more meat than one small band could consumer alone. Members of nearby bands participated equally in the bounty, happily sharing the hunters' success. Survival resulted from the fitness of the group.

Modern social network science has revealed some principles of integrity and effectiveness of social groups that resonate with my understanding of ' social relations and operational roles on the hunt. The network of relations among members of their band reflects some important findings that have emerged from recent social network science experiments. Viable network structures are the evolutionary heritage of humans. However, some network structures work better than others.

Networks in Nature and Society

Societies hold it together in different ways under various conditions. However, when you get right down to the basics, certain principles apply to any group seeking to survive. Industrial societies have taken those principles about as far as they can go. Coordinating operational functions is important but often overlooked.

The COVID pandemic of 2020–2023 demonstrated how the complicated breakdown of international supply chains can severely disrupt the operations of industrial civilization when faced with an unanticipated environmental force. Neither government nor corporate officials had planned for the failure of global dependencies. Many had touted the just-in-time supply of materials for production as the new optimization of the extraction-transportation-production process in modern industries where Nature was assumed to play no role. Then the COVID-19 pandemic disrupted that insular process.

Whatever the scale, any group must maintain a significant level of social cohesion for it to work at all. At large scales, that is difficult in itself to achieve. Yet, it became virtually impossible during the pandemic. A small band eking out a living directly from a harsh environment is quite different from a global-industrial civilization with its complex, large-scale interdependencies. Yet, there is always a need for a good deal of cooperation for any group or organization to flourish at whatever scale it operates. How is that cooperation possible, especially on a very large scale?

In the context of its values of individualism and materialism, cooperation is a highly underrated value in the culture of industrial consumerism—unlike compliance. Some argue that competition for food and other survival requirements is what drives human behavior and that the same competition drives the behavior of modern people in organizations. However, Darwin's concept of survival of the fittest often means the fittest social group, not just each individual, which does not require competition within the group. In complex modern hierarchical organizations, authorities must impose cooperative practices to overcome the destabilizing influence of the dominant culture of competition cultivated by industrial modernism to stimulate consumption in the market and compliance at work.

Jeremy Lent (2021) has demonstrated the clearly inherent qualities of empathy and compassion as primary human characteristics throughout history. The actual lives of our hunter-gatherer ancestors demonstrated these qualities clearly. Few such groups, where cooperation prevails with that level of strength, exist anymore.

Certain things in my academic life have stuck with me as memorable, like Shibutani's lectures or my conversation with Carlos Castaneda in the early 1970s. Castaneda's seemingly magical adventures in the Sonora Desert in Northern Mexico, reported in his first book, *The Teachings of Don Juan: A Yaqui Way of Knowledge,* and several others that followed it had become very popular, despite the fact that it had been his master's thesis in Anthropology at UCLA. Several of his subsequent books reached bestseller status, yet caused great controversy among social scientists, many of whom challenged the veracity of his stories. Castaneda was a great storyteller, but the line between observation and fiction was never quite clear.

In any case, when Castaneda told me about his methodical recording of behavior, using the coding techniques of dance choreography, I found it to be clever and methodologically innovative. Judgments of his writings ranged from dismissing it as fiction dabbling in magical realism to an almost cult-like acceptance of the mystical elements in his stories. We will never know exactly where the line between observation and imagination lies in Castaneda's stories; he died in 1998.

Other experiences made a lasting impression on me, too, such as a near-death experience on a narrow bridge at the bottom of a valley in central Mexico or a near drowning in a riptide off Padre Island, Texas—but those are other stories. Facing death makes life seem far more vibrant—one of the themes in Castaneda's stories—and may turn one's attention to questions of how humans can make their lives even possible under conditions of extreme change. We can find answers to such questions in the relations of people to one another.

Damon Centola (2021) describes in his book *Change: How to Make Big Things Happen* recent research that shows how social networks with certain structures are far more effective at achieving their goals. They not only adapt to change but also make it more than networks with different structures. Key elements in effective social networks (groups) are the interconnectedness of the members and the trust between the members.

Centola compares what he calls fireworks display networks, which have weak ties, with "fishing net" networks, which have strong ties. The former are hierarchical, with most connections leading to a central node. The latter are distributed; that is, each node connects to several others but connections do not focus on any central node. Communication is more or less evenly distributed in the network. Actually, we can arrange social network structures anywhere on a continuum from extremely centralized (hierarchical) to democratic (distributed). The multi-centered network pattern in the center of Figure 1 represents a mix of hierarchical and democratic sub-networks.

In hierarchical networks, members connect primarily through a central node without distributing their links widely among diverse nodes, which makes trust problematic. Rules drive behavior more than trust. Effective networks involve diverse distributed connections among members. Hierarchical networks have weak centralized communication, ties, and control; distributed networks have strong ties and democratic control. People build the most effective networks on trust and respect, not on a hierarchy of power.

Centola's experimental research focused on two kinds of networks, the centralized hierarchical and the distributed democratic. However, in the real world, networks occur anywhere on a continuum between these two theoretical extremes. Figure 1 below shows both ends and the middle of the range of possible social network configurations. Actual network structure can determine how well a group functions and also how stable and reliable it is. Generally, the more distributed and the more trust, the stronger a network is.

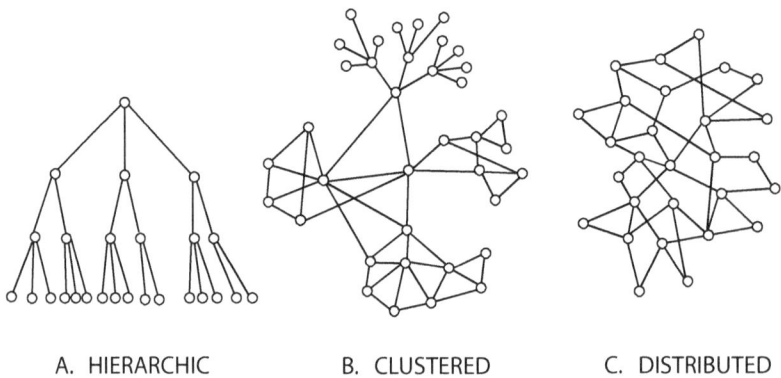

A. HIERARCHIC   B. CLUSTERED   C. DISTRIBUTED

Figure 1. Types of Social Networks

These are just three points on the continuum of network structures. Endless variations are possible. In Network B, for example, the upper cluster is hierarchical, while the other clusters in that network are mostly distributed. In the real world, the possibilities for change differ, depending in part on the configuration of social networks and the level of trust within each, among other things.

This may have profound implications for stability and change in societies and how able they are to respond to changes in their environments that may present threats or opportunities. Hierarchical networks tend to generate increasing power at the top—often detached from the realities of life at the bottom. Instabilities tend to grow throughout a hierarchical network until the power structure breaks down. Distributed networks tend to produce greater wellbeing throughout their structure and, therefore, sustain stability as long as they distribute power and responsibilities widely and fairly. That assures high levels of trust and a sense of fairness in social relations. In hierarchical networks, social control results more from expectations for rewards and losses imposed by the top-down authority.

All this implies some things about how not only small social networks themselves but also how large societies survive or break down and collapse. That also suggests that when we compare our current societal structures, we would do well to consider the destabilization and collapse of other societies.

Hierarchy in Societies

Throughout history, societies have struggled between hierarchical and distributed forms of allocating both power and responsibility in social relations. Authoritarian centralized social structures based on hierarchies of political power have vied with democratic distributed structures based on trust in interpersonal and community relations. In modern societies, that struggle often results from the pressure toward autocratic centralized political power on the part of economic elites. Eventually, instabilities that partly arise from extreme economic and social inequalities in the society's hierarchy result in the rise of fascist tendencies, especially when democratic institutions seem to be failing.

The institutions of representative democracy tend toward failure when they represent powerful interests, not the people. I like to think of it in terms of how human systems work in general. On the one hand, power always offers a means to gain more power. On the other hand, people have always cooperated in seeking mutual goals. They have shared power based on who has the best skills to do one thing and who is more capable of doing another. It is a constant struggle between these two tendencies.

We can find all sorts of complex adaptive systems throughout Nature, including human societies. Ecosystems are an obvious example of complex adaptive systems that self-adjust to changing conditions and self-regulate their internal processes. These keep them within the boundaries of workable relations with other processes within the system. The human body operates that way too. Each living system has a particular structure with some elements of both hierarchy and distributed power, all coordinated in relation to drawing resources and energy from their habitats without disrupting sources of survival.

Ecosystems always involve both competition and cooperation in a variety of ways. They also have self-regulating mechanisms that sustain their stability. Evolutionary ecologist Ruth DeFries (2021) describes these adaptive self-correcting features with the metaphor of a circuit breaker. When a process gets out of hand, the system shuts it down. In biological systems, such stabilizing processes are automatic. However, social rules and norms are not hard-wired. They rely on cognitive decisions and consensus or on some form of force. We must apply the principle of system response to threats and risks to the global political economy today if we expect to avoid global collapse.

When ecosystems thrive, whether in an arboreal forest, a prairie, a marshland, or a barrier reef, we observe that many complex interactions among species of plants and animals exist in balance with each other. Biologists used to talk of the food chain within ecosystems. They now recognize more of the complexity and call it the food web since the structure is distributive in more complex ways than a linear or hierarchical "chain" model can represent.

We all live within ecological systems, whether we know it or not. Ecosystems are complex distributed networks that hold multiple interactions among species in balance, leading to stable network structures. We humans are just one of the multiplicity of species within any ecosystem we inhabit. Because of our technologies of mobility, modern humans move between ecosystems all the time. That is why a pandemic can go global so quickly.

Our industrial economies often upset the balance within and between these ecosystems. Our intense patterns of travel and transport play a large part in disrupting the balance of elements in ecosystems. In fact, the entire Earth System is an extremely complex ecosystem of ecosystems, which we have already destabilized severely at all levels and continue to do so.

Modern industrial civilizations have utterly failed to consider that reality. Neither a suburb nor a city seems part of an ecosystem, but each is, in an unbalanced way—they are more like an invasive species in their habitats. Our merging of technology and energy has allowed us to transcend ecosystem parameters for a while. Now, we are paying the price in terms of having damaged the ecosystems upon which we depend.

Industrial civilizations inherently tend toward hierarchical structures due to the self-amplifying feedback loop of capital accumulation. Those with power (money) can use it to gain more power (money) and limit the power (money) of others. The more money you have, the more likely you are capable of accumulating more. Financial leverage becomes political power.

Medieval European societies exhibited strong tiered structures, made so by the strict hierarchies of the Catholic Church and monarchies with inheritance enforced by royalty's wealth, which enabled them to deploy armies to retain control. The power of investing in industrial development gradually overtook medieval political power as the dominant power in society, eventually rationalizing politics to synchronize with finance capital.

Hierarchy can make the exercise of power efficient in the short run. Efficiency takes a back seat to ruthless exploitation when the power structure is autocratic and authoritarian. This exploitation can target serfs, vassals, slaves, land, and, in the modern context, employees and the public. When hierarchical systems grow large, their increased complexity can, and often does, reduce efficiency, especially when rival factions within the hierarchy struggle to gain relative power. The proclaimed American pluralism is not quite what its promoters claim it to be. Social, political, and economic elites have converged and gained strength to the point where the increasingly rigid structures they dominate become more unstable. The political interests of finance capital have relegated various societal interests to a back seat.

We might describe the US today as a complex hierarchical network of hierarchical networks closer to Structure A in Figure 1 (see page 64). A central hierarchy of financial-political elites dominates the more distributed social networks, such as communities and cultural groups at the bottom of the hierarchy. Political and economic structures have converged into a very large web of political authority and economic power dominated by financial and corporate elites.

Sheldon Wolin (2008) calls this increasingly anti-democratic structure "inverted totalitarianism." Rather than a hierarchy controlled from the top by a political or military dictator, a complex structure of financial and bureaucratic institutions controlled by a ruling elite modulates electoral formalisms. These maintain the appearance of representative democracy. A ruling class consisting of the wealthiest people and their representatives pretending to be democratic leaders occupies diverse positions of authority, exerting social control over most institutions and the public.

The most powerful and wealthy families traditionally constitute the social-economic-political elite. Their children often attend elite universities with "legacy admission" in recognition of parents' financial donations. The schools groom them to inherit positions akin to their parents in the elite institutions of law, finance, government, and industry.

Hierarchical human systems have certain strengths, but they also have weaknesses that tend to grow with time. One is the tendency to distort their economies to the extent that economic policy fails to serve the needs of the whole society. In any healthy economy, money that provides the ability to get things done or to acquire things must circulate widely, thereby engaging the whole population so that the society can flourish. However, the excessive accumulation of money and power among members of the political-economic elite becomes a problem in itself. The billionaire class enforces social control in its own interests, weakening the ability of the larger society to flourish.

On the one hand, differences in access and ability to accumulate money will always exist in an economy of whatever kind. On the other hand, the accumulation of money, like the accumulation of any other kind of power, tends to increase the person or institution's capability to accumulate even more money and power. That is a particularly unstable condition.

In terms of systems science, excessive concentration of wealth is a positive feedback loop that can easily get out of hand in the absence of a countervailing negative feedback loop, such as a progressive income tax system. That is, the accumulation of money is self-amplifying because it enhances a person's power to accumulate more money. Excessive concentration of wealth is detrimental to any economy. Unchecked, it ultimately disrupts the stability of society.

In the 1950s, the US had a strong progressive income tax. Marginal rates (the higher the total income, the higher the rate on any additional income) dampened the effect of extremely high incomes. After politicians eliminated that balancing feedback, income and wealth became increasingly concentrated among the wealthiest classes. This led to the squeezing of the middle class as corporations busted unions and the relative incomes of the majority declined, producing the extreme inequality we see today. Outsourcing jobs to low income countries and the automation of many information-processing and manufacturing functions added to the broad loss of economic participation. Such extreme imbalances make for an unstable society, increasingly unable to protect itself from internal and external threats.

All economies operate within the jurisdiction of some political authority. A society's financial system is typically chartered in some way by the political system to operate in a particular way within prescribed limits. That is, like corporations similarly chartered by states, financial institutions follow regulations, more or less. The degree and kind of financial regulation that exists in a particular nation determines the extent to which inequality of income and wealth exists as a relatively minor variation or becomes extreme, which is a major problem. In recent decades, the financial sector has grown so powerful that it seems almost to regulate political authorities. That is a recipe for societal instability and an increased potential for economic collapse.

Some inequality is inevitable due to the inherent self-amplifying feedback loop of capital accumulation. In this case, "capital" refers to money or assets beyond what is needed for everyday living or ordinary economic operations. However, extreme inequality results from the political failure to regulate the circulation of money and other assets in society. Again, in recent decades, financial elites have increasingly controlled the circulation of money in their own interests, aided by their political agents, adding to the risk of economic instability and societal chaos. Extreme wealth and extreme poverty are two sides of the same coin.

We may find it useful or accurate to talk of the rapacious capitalist class today, the sharks of Wall Street, or the new robber barons emulating those of the late nineteenth century. If political authority does not regulate the economy to sustain a healthy circulation of money, the disparity between the rich and everyone else will continue to be increasingly extreme, with severe consequences.

The concentration of wealth and political power will eventually reach a point where extreme inequality causes the economic system to falter and break down, as did the American economy, beginning in the Great Depression of 1929. The Great Recession resulted from uncontrolled speculation in 2008–2009, when the damage spilled over and into the other "advanced economies" in the interconnected, globalized financial system, causing a near collapse of international finance.

The tendency for a society to move toward increasingly severe inequalities of wealth and income is usually associated with hierarchical political systems—and tends to become more so. Truly democratic systems do not produce extreme economic inequalities. In most modern industrial-consumer societies, the political systems consist of a mixture of hierarchy and democratic formalities. In recent decades, the ratio between these elements has shifted strongly toward increased centralization of political power and the associated growth of the economic power of elites.

Let's face it. The rich are getting richer, the poor are getting poorer, and the middle class is shrinking as it slides into poverty. The old cliché that favoring the super-rich somehow causes all sorts of social benefits to "trickle down" to the less fortunate classes has repeatedly failed to fit the facts. Yet, it persists due to the effectiveness of elitist propaganda. Growing inequality is a demonstrable fact about the contemporary history of industrial-consumer societies, mostly in North America and Europe. The growth of homelessness in the Global North, along with Jeff Bezos' super-yacht and other symbols of absurdly excessive concentration of money, measures well the excessive political and economic power at the top of the societal hierarchy.

The implications of this fact reside somewhere on the far side of unpleasant. As more and more people find themselves in economic despair, the society itself becomes more susceptible to instability and dysfunction. While no society is exempt from this tendency, enough differences exist to make cross-national comparisons useful for understanding the global predicament that industrial civilization faces today.

Every nation has its own unique problems. Nevertheless, on the dimension of inequality in relation to hierarchy and societal well-being, in recent decades the Nordic nations have created distinctly different configurations than the US. American pandering politicians have steadily reduced taxes on giant corporations and extremely wealthy individuals over the past several decades.

In contrast, ever since the 1970s, the Nordic nations have headed in another direction. Typically referred to as social democracies, their political economies distinctly focus on the wellbeing of the population as the primary concern of public policy. US policies do quite the opposite by focusing on enabling economic growth for its own sake. Contrary to all the evidence, the old claim of Econ 101 professors that a rising tide floats all boats still dominates the economic culture. Too many small boats are sinking.

Taxes in the Nordic nations far exceed those in the US, and poverty rates are much lower than in the US, where the rich and powerful pay very little taxes while hiding most of their untaxed riches in offshore tax havens. While people in the Nordic nations grumble about taxes, they appreciate the fact that free, high-quality education is available to everyone. One of the distinguishing features of the culture of Finland, for example, is the widespread high regard for public education and the willingness to pay teachers excellent salaries because teachers are highly valued and respected for their important work enriching the lives of children.

The education and wellbeing of children is a primary value of Nordic societies. Both maternity leaves and paternity leaves exist under the assumption that early attention to the needs of newborns and toddlers is essential to raising children—and, therefore, important for the wellbeing of society. Anu Partanen (2016), in her insightful book, *The Nordic Theory of Everything*, explains this culture of the political priority of human wellbeing in real-life terms. She lived in the US for many years and in her native Finland.

The history of the industrialized nations of the Global North reveals diverse combinations of hierarchy (centralized networks) and democracy (distributed networks) in their structures. In Europe, some nations have retained monarchic structures along with democratic political institutions. Significant traditional aristocratic power remains, both culturally and economically, especially when it comes to ownership of land and industry. Great Britain is the prime example of that seemingly contradictory arrangement, which certainly has its flaws. Until December 31, 2023, when she abdicated, Denmark had a queen, and Sweden and Norway have kings. Yet, their political economies are primarily democratic, although the politics of financial power still prevails.

Hierarchy in and among organizations has dominated throughout most of the Industrial Era. Yet, to the extent that hierarchy, usually in the form of powerfully aligned financial, corporate, and political elites, dominates public policy, people tend to suffer. The elites accumulate more capital and political power, using their increasing resources to acquire even more—unchecked. Such imbalances can only lead to increasing instability and chaos, as is happening, especially in the US and Europe today.

The Functions and Flaws of Hierarchy

If we are to understand and effectively deal with the increasing chaos and destabilization of ecosystems and societies today, it is important to understand how complex systems work when they are stable—and what destabilizes them. In social systems, stability and change have a lot to do with the distribution of power and to the extent to which people consider the existing power structure legitimate and fair—that is, whether or not they trust it.

Most modern people assume that the structure of society must be hierarchical. After all, we see hierarchy in every organization and institution around us. The only issue seems to be how fair and equitable the hierarchy is. The traditional Western family is certainly a hierarchy. Not that long ago, the father-husband "owned" his other family members, as a slave master owned his slaves. He had absolute authority over his wife and children, right up to the right to inflict violence in response to some act of insolence or insubordination, or just because.

While most do not remember the severity of that family hierarchy, plenty of its vestiges remain, most often expressed in domestic violence. Most of us "moderns" prefer a more egalitarian family structure than that. Some parents, in attempting to be fair and equal about everything, go overboard. They treat their children as equals instead of exercising the necessary authority to guide their development, expecting the children to grow up having learned the basic moral principle of living with others. Parental authority is necessary when exercised judiciously with unconditional love.

Many people today consider anthropology to be just one of those esoteric academic disciplines that have no bearing on our modern lives. As an undergraduate, I chose a major in anthropology for a while, before I changed it again, after several other changes, seeking whatever field might be just right for me. I remember sitting in a required college freshman English class when the instructor gave us a homework assignment to write a three-page essay. Three pages? English was definitely not one of the majors I explored with interest.

"How could I possibly think of that much to say?" I thought to myself. However, I had the unrecognized advantage of having suffered through a high school English class under the strict control of a grammarian tyrant. Mrs. Pringle (not her real name) regularly assigned us rather complex sentences to diagram on the large blackboard on the front classroom wall. I didn't realize it until much later, but I had become quite competent at building complex or simple sentences correctly. Nevertheless, I would have to fill in content for that essay assignment. That high school English class has been the epitome of hierarchy. Was it the English-language tyrant or the sentence diagramming that gave me the skills I needed in college? Both. As it turned out, I did have three pages of things to say, although I no longer remember the topic.

That brings us to the question of what the benefits and costs of hierarchy may be. Well, clearly, in many groups someone has to coordinate the members' actions if the group is attempting to complete a moderately complex project that requires coordination of the completion of various tasks to achieve a collective goal. That seems obvious in the context of a football team or a crew on a sailboat, especially for a high-speed aerobatic team. The quarterback must call the touchdown pass; the helmsman must coordinate a starboard tack; the squadron commander must direct the maneuvers.

Nevertheless, must one person really have to have singular authority over the others? Well, that depends on the context. Certainly, that would be necessary in the high school English class where almost none of us students reveled in the prospect of diagramming a sentence in front of twenty of our peers or had any idea of the value of doing so. Yet, in many situations and groups, the members all know intimately how each one of their roles fits with those of the others and with the right motivation they can all work together without some "leader" calling all the shots. They are a team, and that involves trust. The degree of hierarchy needed in a group depends on existing and potential relations among the members and the need for precise coordination. When trust is high, a leader operates more as a coordinator of action others have agreed to instead of as a dictator.

Both the effectiveness of a !Kung hunting party discussed earlier and the findings of recent research in social networks (Centola 2021) demonstrate a general principle of human relations. Success of a group is the product of a generally non-hierarchical network structure in which a society distributes functions and authority instead of centralizing them in a rigid hierarchy. In a distributed network structure, a high degree of trust usually exists between members. That is even more important to members of the Thunderbirds or Blue Angels, for whom trust is paramount. It has become increasingly important for society itself as the threat of converging crises looms larger and larger—right when trust seems to have hit an all-time low.

Whatever the group structure, where a "leader" calls the shots to coordinate an action among the members, trust is paramount. A trusted member with accepted expertise or wisdom for leadership, who has respect earned by action and experience, is not an autocrat. Leadership is a product of trusted relations and the recognition of special talent and skills. Leadership in that sense coordinates distributed authority but does not exercise the authority of a monarch or autocrat. Authority rests with specialized or general wisdom and the trust that entails, not with arbitrary enforced status in a hierarchy of power. It is earned and given, not taken.

Decades of research in organizations, both bureaucratic and less formal, prove that when authority is not arbitrary and hierarchic, work groups perform more effectively. Where authority comes to those responsible, things go much more smoothly and everyone is happier than in a rigid hierarchy where members have responsibilities but no associated authority to make decisions in carrying out those responsibilities based on their own judgment. In some experiments, groups composed of diverse non-experts have come up with strategies for solving problems that are better than those developed by a group of specialists. Diversity breeds creativity; uniformity produces standard responses, which usually miss subtle but important factors.

Unfortunately, power once gained tends to create the path repeatedly walked by the powerful to gain more power. That is why so many groups, organizations, and nations, even formally democratic ones, tend toward increasing hierarchy and centralization, eventually followed by breakdown. Some uneven distribution of authority is probably inevitable, but hierarchy is far less beneficial for everyone than is democracy.

Stability and Chaos in Social Systems

Most folks would agree that the world is increasingly unstable in one way or another, or in several ways. They would undoubtedly disagree on which instability is most important and where the causes of instability lie. There seems to be widespread agreement that people in places of power have abused their power and that it's getting worse. However, in the context of differing ideologies, they usually disagree about what the cause of chaos and destruction may be and which leaders are to blame. Over decades of population surveys, Americans have asserted that the country is going in the wrong direction, whichever party is in office.

Demagogues play on such concerns, trying to appear "populist," while they actually serve their own economic and political interests, which usually align with the interests of the powerful elites about whom the people worry. Hypocritical? Of course, that is what demagogues do. They seek and cause chaos to further their goals by fomenting and manipulating fear to produce resentment and anger, which they try to direct toward their political goals. Anger, amplified across the nation's population produces political violence.

The growth of conspiracy theories is a symptom of the cultural confusion and the insecurities so widely experienced late in the Industrial Era. Conspiratorial imaginaries are just sitting there waiting for demagogues to exploit them by spreading false but persuasive memes. Hate crimes surge. People know that governments and corporations do not always act in the public interest. However, their explanations vary with the extent of their psychological vulnerability to the scare tactics of demagogues, who usually quietly align themselves with the forces of hierarchy and wealth, which they hope to control. Taking power by any means is their core purpose. Creating chaos is their method. They spread social media memes and dark fantasies presented as facts, adding gasoline to the fires of fear, producing more hatred. Antisemitism and racism seem the most persistent and pernicious.

A result of all this is that too many people misperceive the causes of their fears and misdirect their anger toward one scapegoated group or several. The demagogue's focus on demonizing vulnerable groups makes it very difficult to shift the force of public opinion toward policies that improve public wellbeing. It hinders ways to respond effectively to existential threats such as the climate/ecological emergency that most politicians still do not know how to face. That would only weaken the autocratic attempt to seize the highest ground in the political hierarchy.

Meanwhile, our culture allows the super-rich to indulge in riding phallic rockets to nowhere in their competition for the hollow "honor" of displaying their status as the richest men on the planet as they spew vast quantities of carbon into the air below. Symbolically, they can feel superior to the whole world below them by their physical launch to the edge of the atmosphere. Juvenile? At best.

In the meantime, a growing numbers of citizens try to eke out a living, increasingly outside of the global economy of automated growth, or at its edge in corporate retail servitude, or homelessness but always outside any viable ecological system. The institutions and leaders of the globalized system arrange for the dumping of toxic detritus, usually out of sight of all but the most vulnerable victims of the technosphere.

Hunting, gathering, or any productive activity is impossible in today's toxic sacrifice zones found at the edges of active zones in industrial-consumer societies and the Global South. Besides, populations are far too dense to allow such ecological balance. Neither resources nor viable ecosystems exist in or around those sacrifice zones. That is because of the toxic waste from mining, drilling, or processing of the materials and waste of the industrial economy. It happens wherever resources exist or land is needed for refineries or other toxic activities, and residents have no political power.

As often as not, pollution of their land, water, and air damages the health of people whose ancestors lived comfortable indigenous lives there before the pollution began. Usually, they are poor with no political power; they have no recourse to either move or stop the damage. These areas offer no natural sources for living—only "resources" to extract and process for the industrial system—nor do affluent suburbs for that matter. Nothing analogous to a natural ecosystem exists within the sacrifice zones or even in the elite gated neighborhoods or cities of "modernist" political economies. The individual, family, and group find themselves completely dependent on an external economy that may not have made a space for everyone, especially the socially excluded.

Waste from the operations of extractive corporate capitalists, where they extract materials and energy or deposit their toxic waste from both extraction and industrial processes, has destroyed many formerly viable ecosystems around the world. Such operations have disrupted what had been viable social systems as well. Meanwhile, industrial consumers remain personally and culturally detached from the dying ecosystems they inhabit. They stay dependent on the industrial political economy, which is culturally detached from its habitat in the larger Earth System, even as it approaches and passes planetary limits to growth.

Ironically, when the global-corporate-political economy falters, affluent cities and suburbs will have little more basis for survival than the destitute sacrifice zones do now. The supply-chain chaos that accompanied the COVID-19 pandemic was a demonstration in miniature of this impending global failure. Ultimately, establishing a viable ecological civilization will require widespread restoration of degraded habitats everywhere.

Having long since abandoned any sacred or direct reciprocal relations with the Natural Order, we factory, service, and office worker-consumers have briefly enriched ourselves with stuff but at great mutual peril. The proliferation of suburban storage units directly indicates the absurd world of overproduction and overconsumption in which we live. As our economic productivity has grown vast, we have isolated the global economy's most vulnerable victims in sacrifice zones, homeless encampments, refugee camps, detention centers, and prisons, each of which reeks of social, as well as chemical, toxicity.

Our largest and most powerful institutions put growing numbers of people at direct risk every day. Now, they put all of us, even those holding the greatest economic and social resources, in great imminent danger. The fortress-like "gated communities" and elaborate prepper hideouts of the extra-affluent classes ultimately offer no real security, except in the narrow minds of their owners—but not for long. No fortress in history has ever survived without secure supply lines from surrounding hinterlands.

We, humans, have lost our connection to the very Earth System that sustained our flourishing for thousands of years before the advent of fossil-fuel energy and hyper-mobility. No wonder we find such pervasive discontent everywhere as we approach the end of industrial civilization. We desperately need a New Great Transformation of society to overcome the greatest predicament humanity has ever faced, and we need it now.

CHAPTER FOUR

Industrial Modernity
and its Discontents

HUMAN EVOLUTION HAS gone through several stages, each more complex than the last. It has also undergone several transformations in how we engage the world and each other. What we generally call "modern civilization" began, at least in some ways, when societies in Europe started breaking with the traditional social constraints of feudalism.

New forms of economic development resulted from the invention of a wide variety of technological innovations with the explosive growth of science in the seventeenth century following the Renaissance (rebirth) of science, the arts, and culture in Europe. With creative thinking furtively freed from the constraints of church dogma, these developments allowed not only discoveries but also the rationalization of economic production processes of all kinds. That required the rationalization of society as well.

Conveniently, the capital accumulated by conquest and colonial exploitation around the world was available for investment in new ventures. However, implementation of the new production processes required severe societal changes. Move fast and break things is not just Mark Zuckerberg's predatory motto for Facebook; the idea is as old as Western culture. Industrialization has broken traditional societies, and unfortunately, Nature's ecosystems as well.

The most common term used to describe these developments was "progress." However, there were downsides to all this, very little of which were recognized as such at the time. Ultimately, the economic "progress" of what we now call the Global North had rested on some pretty nasty practices, as certain European nations learned how to navigate and use firearms to explore and exploit the rest of the world. They used those technologies to dominate and, in too many cases, exterminate the peoples they encountered around the world. A new form of colonization of their own populations followed.

Industrialization came only after the Age of Exploration (and exploitation), when Portugal, Spain, England, Holland, and others established colonies and opened up trade and expropriated materials. They both exploited and exterminated the peoples of diverse societies of the Global South. The subjugation of much of the rest of the world began when explorers conquered and colonized peoples wherever there were riches available for the taking. This resulted in the accumulation of wealth by both theft and trade. Plunder produced capital, which the increasingly wealthy nations of Europe then deployed to establish and expand new industries and markets, which relied on the plethora of inventions associated with the new science of post-feudal Europe.

Christopher Columbus and his Spanish expeditionary forces committed barbarous genocide upon the peoples he discovered when he first made landfall in what we now know as the Bahamas, then Cuba, then Hispaniola (now Haiti and the Dominican Republic). Today, in a new recognition of Columbus' victims, some cities and states in the US celebrate Indigenous Peoples Day instead of the traditional Columbus Day. The Eurocentric notion that Columbus "discovered" the "New World" reflected the presumption of racial and cultural superiority of the West (Global North), blind to the legitimate existence of its victims. This continues even to this day in the racism and White nationalism that not only persists but surges in Europe and the US as industrial civilization falters. The dehumanization of colonized peoples also reminds me of Vladimir Putin's defining Ukraine as not a "real nation." Blaming the victim is not new.

For scholars of colonization today, Columbus is no longer a hero of "discovery" and "progress." Many historians now recognize that the names of holidays, cities, mountains, and rivers named for these explorers and colonizers represent men who led, at best, questionable lives of plunder, exploitation, and brutality. The slave trade was one of the later episodes of persecution and degradation of other humans defined as less than human by the new inhabitants of the "discovered" lands seen as ripe for expropriation. If non-Christian, not quite human (for example, not-quite-White) "savages" inhabited a land, then the explorers and colonists defined the land as essentially uninhabited and, therefore, free for the taking.

It is striking to note that the treatment of European populations by the elites deploying the industrialization process parallels the treatment of indigenous peoples of the Americas, Africa, and Asia by their European conquerors during the previous centuries. The enclosures of traditional small agrarian plots along with the clearances of the peasants from their traditional lands and the commons they had shared for centuries in England and Scotland destroyed communities and forced their residents into towns and cities where the new industries were forming. They became wage slaves.

In several ways, the processes of enclosure and clearance continue today, only in different forms. Technological and organizational "progress" entails a variety of causes of discontent for millions of people just trying to "make a living" in an environment over which they have little or no control. India is a prime example of modern enclosures and clearances in which corporations (with legislative support) force small farmers off their land, replacing traditional ecological agriculture with genetically modified monocrops laced with petroleum based fertilizers, insecticides, and herbicides for global trade. Livelihoods and local food are replaced by corporate profits (Shiva 2015). The people continue to lose whatever control that remains in their own lives.

The New Great Transformation and the Old

Today, as we traverse the last stage of the Industrial Era, we can look back on the sequence of developments that, with severe disruptions of existing social relations, produced the most powerful system that ever exploited the planet and its people. It is not over yet. That power is a two-edged sword. With one edge, industrial civilization provided an affluent lifestyle that most would have considered unimaginable before the twentieth century. For a time, that included an expanding middle class, the members of which believed that everyone would eventually participate in affluent consumerism. However, it did not last.

With the other edge, industrial modernism forced the vast majority of the Earth's peoples into deep poverty. Many are destitute to the point of starvation. At the same time, the middle classes of industrial nations have been shrinking for decades, breeding growing discontent. With the deepening climate and ecological disturbances, not only has the industrial modernization project begun to falter, but now gaping disparities of income and wealth are accelerating while risks of widespread food insecurity spread with accelerating climate chaos.

Despite protestations by the privileged members of the mainstream (neoliberal) economics profession, the politicians, and the financial and corporate elites they serve, who claim more economic growth will resolve these gross inequalities, the facts speak a very different language. The Earth System has already spoken clearly, declaring its objective limits to the growth of the global-industrial-political economy. Industrial civilization is now confronted by the most vexing predicament humanity has ever faced—and has yet to face. The Earth System has boundaries, and we are on a phantom path toward an illusory great prosperity that we will not find. Ideology and political power cannot alter physics and biology in the real world.

Why is this tragedy unfolding? Because the basis of the industrial-consumer-political economy of endless growth is a principle that worked (for some) under the conditions that existed when it began, but boundaries abide while conditions on the ground have radically changed. The Industrial Age has outgrown its habitat, and there is nowhere else to go. What once seemed true no longer is for two reasons.

First, long before anyone could (theoretically) realize a global-industrial-consumer economy for the planet's people, both the global climate and many of the biosphere's ecosystems had already begun their collapse into chaos. This has quickly led to a great acceleration of species extinctions—and even possibly toward that of our own. Although the limits to growth are already being felt, the distribution of that growth is extremely uneven. The disparities between the wealthy and the poor have never been this extreme, both within and between nations, and they are becoming so extreme that they portend growing societal chaos followed by systemic collapse.

Second, even if the climate/ecological emergency were not as severe as all the evidence demonstrates, the Earth System would need resources equivalent to several additional planets for Earth's peoples to sustain the current level of conspicuous consumption and waste seen in the Global North alone. The math is simple and clear. It just does not add up.

The immediate emergency requires a New Great Transformation of human society itself, far greater than the Industrial Revolution caused, to slow if not stop the severe disruption of climate and ecosystems toward which we career. Ironically, a new transformation of industrial societies has already begun and not in a good way nor one of our choosing. Yet, our indiscriminate disruption of natural Earth-System processes triggered it. The current trajectory of industrial civilization staggers toward collapse.

We can attribute all of this to a fundamental, yet understandable, cultural flaw. Our forefathers built this industrial civilization on a partial truth, a temporary truth that becomes a falsehood when taken too far. Indeed, we have already reached the point where that truth has become a lie. The premise of the global political economy is the necessity of perpetual economic growth on a finite planet, and that worked for those who ran it for as long as there was room for further expansion. There is no more room. In ecological terms, the expanding global-industrial-consumer economy has already exceeded the carrying capacity of the Earth System.

To understand this well enough to make the right decisions now, we must reflect on the great transformation that was the Industrial Revolution itself, which put us onto this path. In 1944, Karl Polanyi published a brilliant book, *The Great Transformation,* which was entirely out of step with the societal trends of the time. A great acceleration of economic development was about to follow the Second World War. Polanyi's book pointed out the unresolved problems with unbounded industrial development. However, his book only sold a few hundred copies at the time and then languished in relative obscurity for decades. It contradicted the ideology of economic expansion—the presumptive engine of human progress.

Karl Polanyi explained how the Industrial Revolution changed everything in the societies it touched. Even so, the ecological consequences of the globalized techno-industrial system have reached far beyond anything Polanyi could have imagined in 1944. He proposed that the Industrial Revolution was a great transformation unlike any change humanity had undergone before. He was certainly right about the Industrial Revolution and its disruption of traditional societies and communities, but he could not have imagined that a great transformation would unhinge the Industrial Age, the initial problems of which he understood so well. Yet, here we are, facing the imminent collapse of the ecosystems upon which we depend for survival.

By a great transformation, I mean two things. First, the two hundred years or so of "industrial progress" have so disrupted the entire Earth System that the conditions for many life forms on the planet put them into immediate jeopardy—a great transformation emerging. Second, the Industrial Era has also caused fundamental changes in the structure of society itself, resulting in an extreme hierarchy of social control from a financialized political economy, which has come to dominate society itself. We cannot yet determine the outcome. It depends on whether or not, and to what extent, we humans respond effectively to subdue the disruptive Earth-System changes we have caused—a great transformation either way.

The New Great Transformation of society is well underway. However, its ultimate direction remains unknown. Either it will continue on the path toward societal collapse or it may resolve into the transformation of the existing global hierarchy into an ecological civilization—an admitted longshot. The determination of that direction depends entirely on human awareness and, above all, radically different human actions.

The Industrial Revolution literally revolutionized the social order by imposing a new form of control over traditional societies. No longer would natural social relations in communities determine the way people lived. The organizational imperatives of industrialization included corporate access to detached workers unconstrained by personal, family, or community obligations. In important ways, social control shifted from family and community to the new economic system organized around industrial production. That production served international trade, such as wool fabrics manufactured in British factories for trade with other nations. It did not serve the local populations whose communities it destroyed. Former self-sufficient agrarians became wage laborers who struggled to earn enough to subsist.

The multifaceted great transformation confronting us today results from the "progress" of the industrial project itself. We depend on the Earth System, our habitat, yet we have severely disrupted the stability of that habitat, which had lasted over the eleven thousand years of the Holocene Geological Epoch. Extractive and industrial-consumer processes spew vast quantities of toxic chemicals and industrial waste into living environments. Micro-plastic particles even invade our bloodstream, entering our organs and brains. These incursions have destabilized the many complex living systems, both species and the ecosystems of which they are members.

Deeply disturbing changes throughout the Earth System are well underway, and we must respond quickly. An adequate response cannot happen within the existing global political-economic hierarchy. The only viable strategy must involve both science and core human values, even ancient wisdom, not simplistic illusions of technological fixes, lest we fall into unrelenting chaos. That means overriding the forces of hierarchy, a daunting task.

The values we need to drive the transformation in the direction of an ecological civilization directly contradict the culture of endless economic growth. The political and cultural obstacles to formulating effective responses are enormous, yet must be overcome soon. Political "leaders," from mayors to the president, continue to talk of growth as if it were the sole measure of human progress. In fact, it has reached a dead end.

Having already entered the current transformation of our relations with planet Earth, we must not only become aware but also take actions to gain control of our role in that transformation to bend its trajectory in a survivable direction. Otherwise, we will continue on a downward spiral toward societal collapse. That is the most difficult thing to imagine, for we have no comparable historical model to guide us. Collective creativity is the watchword of our future.

Polanyi's book *The Great Transformation: The Political and Economic Origins of Our Time* not only explained the origin and nature of the Industrial Revolution in great detail. It also warned of the difficulties it posed for societies, even as it provided a new economic flourishing for the societal hierarchies in which it occurred. Polanyi put a great deal of effort into trying to resolve the societal disruptions and exclusions caused by industrialization, even as it offered an explosion of economic opportunity for some. He even warned of the potential for ecological damage resulting from industrialization. He could not have imagined how bad that would really get.

The history of cultural evolution in Europe led directly to the ideology of Earth plunder and perpetual economic growth, as well as the social and ecological consequences that it entailed. Religious and cultural beliefs of medieval and then Renaissance society envisioned "Man" (not woman) as God's steward of the Earth, separate and superior, charged with the task of utilizing its resources for human needs. Men viewed themselves as intellectually and spiritually separate from Nature.

The emergence of the new science from Renaissance thinking and experimentation produced many innovations and basic knowledge. The resulting technologies yielded a much greater advantage in exploiting the physical world than ever before thought possible. However, its application to resource extraction, economic production, and commerce caused huge, although poorly understood, ecological and societal changes. We have now entered a new era of extreme change, which will end in ruin without severe human intervention against our continuing collective folly.

Discontents and Disorientation of a Rational Society

The Industrial Revolution was a top-down process that enforced a new kind of hierarchy. Peasant society continued along its traditional path until forced to do otherwise. As industrialization accelerated, investments by aristocrats and merchants in new industrial techniques required large plots of land and "free labor"—workers unattached to traditional families and communities—to work in factories and less so in industrializing agriculture.

Traditional societies embedded economic activities within the traditional cultural practices of the people so that people integrated economic activity into core cultural values. Industrialization changed all that. The new industrializing economy embedded society within it and organized it to achieve ever more rapid industrialization. Inevitably, this ultimately destructive transformation of traditional communities caused massive dislocation of individuals. It destroyed their communities because they did not fit the model of the new economy. Individuals became isolated from the mutual support structures that their traditional communities had sustained because those structures broke down under the enclosures and clearances. The industrial modernity we experience today retains and amplifies that isolation, even as we imagine that suburban developments are "communities."

The traditional small agrarian plots from which peasants had drawn a subsistence living for centuries endured by making most tools and products they needed. That ended when the owners of the land—the lords—enclosed the small plots, combining them into larger tracts so that they could operate the new implements of industrial agriculture more efficiently. The commons were simply lost as the owners cleared people from the land. The commons became part of the same larger acreage in the process. The enclosures and clearances changed English and Scottish societies in the most fundamental ways before similar changes spread across Europe and America in different ways.

Around the turn of the twentieth century, Max Weber, a student of political economy who became one of the founding fathers of modern sociology, studied the core processes and structures of societies and their institutions. He used comparative methods that would later gain central importance for the social sciences. Weber developed several key concepts for understanding the differences he saw between societies. He was particularly interested in how different forms of authority worked. But his key concern was the new forms of social organization unfolding before his eyes as European societies industrialized.

Authority, distributed either broadly or centralized, is the keystone in any system of social control. Bands of hunter-gatherers distribute authority based on specialized roles, each of which entail special skills, such as hunter, chief, and shaman. Industrial societies involve a great deal of coordination of people specializing in a multiplicity of roles in very complex organizations, mostly corporate and government. Unlike the hunter-gatherers, in each of those organizations control operates through a centralized hierarchy of formal authority—the regulatory structure of the organization.

Rigid hierarchies of control tend to demean people—who generally feel like cogs in a machine—and treat them as mere commodities. The rationalization of organizational control depersonalizes interpersonal relations in a bureaucracy. That leads the person to feel detached from the organization and other persons within it. Nevertheless, most people look for ways to make their relations personal. In a bureaucracy, that is a two-edged sword.

Even today, over a hundred years after Weber explored the widespread disenchantment that seemed to accompany the rationalization of society, a good deal of controversy continues around how to "administer an organization to humanize" it. It reminds me of Charlie Chaplin in the film *Modern Times*, trapped in the cogs of a giant machine, unable to escape its dehumanizing perpetual operations.

One of the most powerful proposals for humanizing an organization is to rearrange the hierarchy so that, to whatever extent a person has responsibility for a function, they also have the authority to make decisions about how they will execute those functions. Hierarchies resist such reforms.

Owners and executives in complex organizations tend to resist distributing authority because, aside from just wanting to keep as much power for themselves as possible, they do not trust the workers whom they underpay and exploit. Why? In most organizations, most individual's involvement is limited to performing labor for pay. To see the contrast in workers relations with the corporation they work for, observe Costco workers and Walmart workers, look at their pay scales, scheduling, etc., and you will see the difference.

What is the stake of a worker in the organization that employs them, even when the executives try to portray the organization as a "family"? Who do they trust? They trust only a very few individuals with whom a personal bond exists—but not because of their organizational roles.

Most workers know that even companies that have maintained pension funds for their employees have often simply taken the money saved up for worker pensions and used it for other purposes, such as when mergers and acquisitions call for liquidating corporate assets. From the perspective of the capitalist, this is merely "rational" behavior in seeking to maximize profits and capital expansion for the firm or its owners. For the most part, workers' wellbeing is not a consideration in that rational calculus.

In the relatively few worker-owned corporations in the world today, things are different. Workers sit on boards of directors and contribute to policy decisions; they have ownership stakes in the company. The most cited example (probably because it is the largest and most successful in the world) is the Mondragon Federation of

Cooperatives formed in the Basque region of Spain. Its worker-owned cooperatives operate many businesses in Europe. Its workers own it entirely, contradicting the elitist mentality of many corporate executives in the US, for whom workers are mere resources, like steel or plastic, but are seen as having much less value.

At the other end of the size range, each one of the !Kung tribesmen on a hunt in the Kalahari Desert in southern Africa before the Europeans came in and forced them out, had an equal stake in the success of the hunt and the survival of the whole group. Modern organizations have not replicated the mutuality of interest and consequent commitment of each member except to a small degree by a few progressive thinking organizations today. The rationalization of organizations in industrial societies has produced growing disenchantment with the world and detachment of individual members of organizations from the corporate culture.

On a broad scale, this produces not only disenchantment with the world but also deep discontent with one's personal life and resentment of the society that produced it. Societal instability emerges in political frustration, resentment, fear, and hatred, especially when egged on by demagogues running for office and by social media influencers spewing hate.

These sociopaths gain attention by fomenting as much resentment and hatred of the usual suspects, vulnerable groups such as Blacks, Mexicans, Jews, Muslims, Asians, other immigrants, etc.—further disturbing the tenuous social order. Continuation of such instabilities and hostilities usually leads to political violence and personal despair often expressed in domestic violence.

If you consider these sorts of consequences of what we call a "rational social order," the question arises as to what is rational. At the turn of the twentieth century, Max Weber looked at the "rationalization" of society as the form of authority that most efficiently organized the new industrial economy. It was efficient, as far as it went. However, "rational" meant to optimize the goals specified top-down by the owners and managers of an organization. Wellbeing of workers is not so often high on the list of goals. (I know, it can conflict with maximizing "shareholder value.")

Consider for a moment the other forms of authority that Weber observed in various societies (1947). Traditional authority existed in societies where social relations evolved gradually over time and social roles were clearly defined. Pretty much everyone was in agreement about, or at least accepting of, the social order as part of a broad cultural consensus as to how society should operate and how people should behave. After all, in traditional societies, virtually no alternatives were present.

I should point out that Weber used the method of constructing "ideal types," which conceptually capture the essence of a particular form of social action. So, traditional authority operated primarily based on cultural consensus. However, no type of authority perfectly fits Weber's ideal types. Even where a society operates on a high degree of consensus, there will be rebels who seek another path.

Sometimes a charismatic leader will emerge with high ambitions that do not fit within the confines of traditional precepts, a rebel who envisions another path that conflicts with existing authority. Charismatic authority is personal. Its legitimacy results from some perceived personal characteristics of the leader, not from rational-legal rules or traditional values. Charismatic authority is neither rational nor traditional; it is, indeed, personal. Two examples of politicians, who could not be more different from each other, are Mahatma Gandhi and Donald Trump, which is not to say that these two men shared any other characteristics.

Sometimes charismatic leadership will result in a social movement or a revolution that overthrows the traditional or even rational-legal authority and places the charismatic leader into power but not always. There are plenty of examples of political crises going either way. The power of charisma will likely go nowhere without a base of discontent among the people. Contentment does not lead to revolution. Remember, an ideal type is a simplification of a very complex world. It does not suggest that a particular situation, person, or outcome is either good or bad. Donald Trump was a charismatic leader whom many would call a demagogue since his motives and actions seemed so narcissistic and sociopathic. (Was that past tense "was" just wishful thinking?) Yet, he was able to move a significant number of discontented voters and even foment a failed insurrection and more.

When authority is traditional, a high degree of cultural consensus is involved—even when the social structure is relatively hierarchical— which means that individuals internalize the legitimacy of authority. Personal values and social rules are consistent, and the social order is stable. Under conditions of cultural ambiguity and economic change, things can get quite unstable as people look for certainty and too often find it in the claims of demagogues.

The stability of traditional, especially folk societies, changed radically with the advent of industrialization. With the rise of capitalism and the "rational" organizations focused on economic production and capital accumulation, the world became secularized. That is, the rational order of modern organizations enacted values and beliefs about the world that were no longer grounded in religious or cultural precepts, as the organization of traditional societies had been.

Instead, the culture of modernity founded its core beliefs on ideas about the material relations of economic production itself, even though religious and other cultural values persisted to some degree. That is why Weber talked of disenchantment with the world—ideas about human life were no longer primarily grounded in religious, mystical, or magical ideas or even community values. They focused on the material relations of humans to the industrial and related organizational processes that were taking over the world.

With industrialization, rational-legal institutions took over most major social processes, leaving many folks disenchanted with the somewhat mystical world of traditional values. Even so, the entrepreneurial spirit found its justification in certain underlying assumptions of protestant Christianity. Put simply, pious merchants saw their economic success as a preordained ticket to heaven. Thus, people retained religion within the quest for material accumulation (Weber 1905).

Of course, a moral distinction between rich and poor could not lead to social contentment. For many, I am sure, the idea that wealth implied moral superiority was quite a stretch. After all, many poor folks knew all too well how their "betters" came by their riches, whether by smuggling or the practice of slavery, etc. Besides, the many negative impacts of industrialization on traditional society have never been resolved. Even by 1944, Polanyi, who understood industrialization better than anyone, could not find ways to ameliorate the forces that excluded so many from the benefits of industrialization. Discontent was sure to follow and remains with us even more strongly today.

The Roosevelt administration had deployed his New Deal policies and programs in response to the Great Depression of the 1930s in an attempt to save capitalism from itself, which capitalists bitterly opposed. Only the economic ramp-up for World War II enabled a return to full employment after the New Deal petered out with the 1937 recession that critics had likened to the 1929–1930 depression. Today, liberals still struggle to save capitalism from itself (Reich 2015), just as most try to solve the climate crisis within the system that caused it.

The great acceleration of economic life in the wake of heightened wartime manufacturing capacity and plentiful energy production assured new opportunity for many—if you don't count Black folks,

Mexican- and Asian-Americans, and other "others" who, if they had new opportunities, experienced them only at the bottom of the social-economic ladder.

In any case, every cycle of economic acceleration has its end. The illusion of endless economic growth will also end, if not before growth itself. Discontent is just as undeniable as the decline of industrial-consumer affluence that grows more apparent every day.

Growth & Opportunity Become Dead Ends

The eco-modernists imagine that economic growth can perpetuate through an imaginary decoupling of environmental pollution from a modern industrial economy, assisted by innovations such as advances in technology and the discovery, or creation, of new materials. Envisioning a material world as unbounded as their sci-fi imaginations, they claim we can overcome any limits to growth, such as peak oil followed by a steady decline in production or the exhaustion of various minerals and other resources.

However, the evidence does not justify their faith in industrial technology. There are only so many elements in the periodic table of chemical elements and only so many sources of energy. Materials science must work within the limits of physics and chemistry and within the limits of the subsystems of the Earth System itself. Limits may be denied, but they abide.

Mainstream economists have long held an optimistic view of human progress through economic growth since before the Industrial Revolution and right up to today. Even as diverse Earth-System components destabilize and threaten the very foundations of human survival, these economists persist with their peculiar self-contradictory theories of endless economic growth. Of course, there is a reason for this, but it is a dream, a utopian imaginary not based in any economic facts or on-the-ground scientific evidence. Conventional or mainstream economics is really more of a belief system, an ideology, than it is a science. Economic and political interests, not science, drive it.

Science does not ignore evidence; it carefully examines the facts seeking patterns and predictability. If one scientist ignores certain evidence, another is happy to point it out. Scientists try to understand and explain the facts. One fact is so simple, yet profound, that the evidence is as much a matter of logic as it is a matter of obvious facts. We unequivocally know that planet Earth, being a finite object with its specific size and composition, possesses limited resources. Yet, mainstream economists consistently ignore that simple, but obviously profound, fact. That is not surprising since it directly contradicts their entire worldview. They and their political masters persist in asserting the illusion that economic growth can continue, and must, indefinitely. Yet, the limits to growth are palpably very real.

A young team of systems scientists and graduate students at MIT developed the first major computer models of economic growth and its material consequences around the turn of the 1970s. They wanted to project probable changes in resources and their availability for populations, and the Earth itself. They used actual data on extraction, production, consumption, and waste and then projected the data trends over subsequent decades under different assumptions to learn where the trends might go.

The computer simulations of rates of resource extraction, industrial production, consumption, and waste that the MIT scientists led by Donella Meadows had developed turned out to be remarkably accurate. Their models forecast the global depletion of resources under conditions of continued economic growth into the early decades of the twenty-first century (Meadows et al. 1972). Despite their denigration and dismissal by mainstream economists, the MIT forecasts turned out to be quite consistent with subsequent events as they have unfolded even right up to the present. Nevertheless, economists and politicians continue to ignore or deny the facts in favor of their utopian dreams.

Despite mounting evidence, the idea of environmental or material limits to human innovation and economic growth gained little credibility among economists or politicians during the decades of strong growth in technology and industry in the second half of the twentieth century. The landmark study "The Limits to Growth" (Meadows et al. 1972), which had predicted that resource shortages would begin within less than a half-century, was widely dismissed by technologists, futurists, and economists alike. Yet, its forecasts proved remarkably accurate.

William Catton (1980, 2008) and his colleagues, including Riley Dunlap, carved out a moderate niche to explore environmental factors within American sociology. Yet, despite the environmental implications of their research for the limits to growth and their impact on society, their work remained separate from most sociological research or theory and was virtually unknown to many policy analysts. That ignorance and indifference parallels the denial of climate science in political discourse today despite explicit warnings by climate and ecological scientists. Those who should have known better ignored or dismissed renowned NASA climate scientist James Hansen's testimony before Congress in 1988. Hansen explicitly warned of the impending climate emergency caused by industrialization. However, knowing is not enough; one also has to both face reality and care.

Our cultural hubris matches the material overshoot of the global political economy. The fact of overshooting is not new. People experience it every day. Sometimes we spend more than we earn, resulting in unmanageable debt or even financial collapse and bankruptcy. We may drive too fast on an icy road and overshoot the stop sign as we slide into the intersection, risking a sudden crash into cross-traffic.

Yet, many people have a great deal of trouble grasping the idea of overshoot on a larger scale. It seems that awareness of the potential for rapid ruin that Ugo Bardi (2017) calls the Seneca effect has found little or no legitimacy in the consciousness of industrial-consumer societies, even as that very threat looms large and near. Figure 2 below represents gradual growth to the point of overshoot and then rapid decline as a system collapses. This is the typical pattern of the growth and decline of empires.

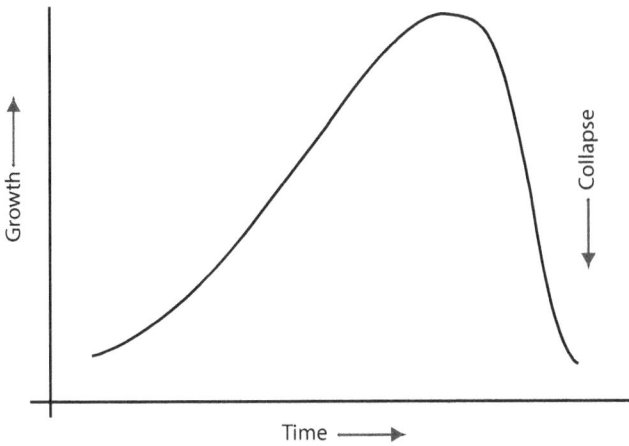

Figure 2. The Seneca Effect. Slow Growth, Rapid Ruin, (After Bardi, 2017)

A key concept related to overshoot, derived from ecological studies of species in their habitats, is carrying capacity. In population biology, carrying capacity is the maximum number of a particular species that a habitat can sustain indefinitely and still keep the ecosystem in balance.

When a species stays within the carrying capacity of its habitat, it remains in balance with other species of flora and fauna and with the material conditions in that habitat, which sustain a delicate ecosystem stability that evolved over a very long time. However, if a species grows out of balance with the other parameters of its habitat, that is, when it overshoots the carrying capacity, a rather sudden decline in its population will likely occur. I watch this principle play out in the high desert of Northern New Mexico every year as the populations of rabbits and coyotes ebb and flow in lagged synchronization with each other, as annual variations occur in rainfall and available food.

With a good snowmelt, the rabbit population expands rapidly with the increased food supply that becomes available in the spring. The abundance of rabbits allows the coyote population (their primary predator) to grow. Subsequently, the rabbit population declines due to increased coyote predation. The extent of these changes depends on environmental conditions, such as precipitation, vegetation, and the presence of other predators, such as bobcats or human hunters. Seasonal oscillations of rabbit and coyote populations are common, but they usually stay within a stable range.

Over time, without external interventions such as gun-shop-sponsored coyote hunting contests, populations tend to return to sustainable levels, reestablishing ecological balance with modest fluctuations. However, if the overshoot of a species destabilizes species relationships within the ecosystem of their habitat beyond a tipping point, the population of a particular species may very well collapse with unpredictable effects. The extermination of wolves in Yellowstone National Park indirectly caused damage to Aspen groves due to expanded elk herds. With the reintroduction of the wolves, Aspen health returned.

Humanity today has reached planetary overshoot, having expanded both our populations and consumption beyond the planet's capacity to carry the load of our global-growth economy. Unfortunately, it looks like the global-industrial-consumer political economy and the populations that depend on it will continue to grow until environmental conditions force their collapse. When a population overshoots the carrying capacity of its habitat, that population collapses. The ecological balance may not recover for a very long time, if at all. Already, humans and domesticated mammals far outnumber the total of other mammal species in the wild. We have come very far from the ecological balance among wild species.

Unfortunately, there has been little in the way of human self-regulation in our relations to our ecosystems beyond pious false commitments to meet carbon emissions or global temperature "targets" by such and such a date. Linear thinking within the frame of finance capital supersedes any other perspective. And no significant action seems likely since the financial and political elites avoid even acknowledging the overshoot or its dangers. Yet, fear is a powerful motivator. The growing intensity and frequency of destructive super-storms, droughts, atmospheric rivers, floods, sea rise, and agricultural failures draw increased attention from news media, which also steadfastly ignore human causes.

Unless we restrain economic growth and institute a redistribution of wealth in a transition to a wellbeing-based economy, unrestrained growth will result in massive societal and ecological collapse. The alternative is to create the new ecologically tolerable political economy that society needs to survive. Because of its inherent need to grow, the global-neoliberal-political economy has no mechanism for achieving balance with its habitat, planet Earth. That will only happen if humans intervene to impose unprecedented contraction, redistribution, and transformation.

We have not even publicly discussed this, no less taken real material action, and no institutional mechanism exists to do so. Nor have we seriously considered the actual level of societal change necessary to come close to adequate reductions of resource and energy use, pollution, and waste. That is one of the main reasons I am writing this book. Adding renewable energy production to feed existing overconsumption is no solution.

In stark contrast with necessity, the continuing global-industrial overshoot has already begun to change the entire Earth System permanently in extreme and unpredictable ways. Chances for any rebalancing seem increasingly remote. That is why many geoscientists have concluded that we have entered a new Anthropocene Geological Epoch, defined by the human impact on the planet. We humans have already changed the Earth System in likely unrecoverable ways. The stability of over eleven thousand years of the Holocene Epoch is effectively gone, caused primarily by the deployment of industrial capital to extract financial value from and pollute the Earth System.

The bottom line is that the globalized industrial economy is unsustainable and will collapse far sooner than most expect since the already overburdened global biosphere can no longer support it. If the collapse of other complex systems is any indicator, ruin will be rapid. The simple mathematical fact of the imbalance between the exponential trajectory of conventional economic growth and the demonstrably finite limits of the Earth System in which it operates leads to only one conclusion.

We have already come very close to the tipping point where only a New Great Transformation of human societies can avoid approaching a catastrophic dead end. Yet, to assume that nothing can be done is a grave error and a self-indulgent psychological escape mechanism unbefitting any sentient being.

CHAPTER FIVE

Consequences of Modern Social Structure

HISTORY IS REPLETE with examples of both hierarchical and distributed systems of social control, each of which produced its unique social order. The seemingly endless struggle between various forms of authoritarian and democratic political systems continues today. However, the trend from before medieval times to the present global-industrial-political economy has moved away from relatively simple systems of authority, both hierarchical and egalitarian, to structures that are far more complex. This shift occurred as technologies for dealing with Nature for survival, profit, and social power became ever more sophisticated. The Industrial Revolution took the complexity of authority in society to a completely new level of hierarchical power.

It is important to remember in this context that authority means *legitimate* power. Social control exercised with legitimacy by whatever political structure is in place is what we call authority. On top of that, we must consider the informal kinds of power that often influence formal authorities. These include lobbyists who pay for the decisions of politicians, litigants who appeal to the courts for special status, or the personal relationships that develop at work to achieve personal goals but may interfere with organizational policies or rules.

Innocent interpersonal relations, corrupt backroom deals, and other kinds of informal influence can change the character of an organization and result in embezzlement or other violations of norms or laws, even to the extent of forcing corporate bankruptcy—all of which undermine legitimacy. Informal power can even result in laws that give power to certain entities, like corporations, increasing their position in a hierarchy of control. The possibilities are endless, limited only by the actual enforcement of laws and rules and the extent to which hierarchy can sustain its legitimacy. In the most extreme cases, such corruption of authority may be pervasive. In the aftermath of the September 2023 floods that devastated Derna, Libya, citizens demanded that outside companies do any reconstruction. They perceived the entire political economy as riddled with corruption, which they blamed for the failure to maintain the dams that collapsed in the heavy rains.

In the Citizens United case brought to the US Supreme Court on January 21, 2010, the Court ruled that corporations have the same rights as individuals to pursue political goals through "free speech," in the form of money spent on political causes. This ruling favored the interests of corporations. The problem, of course, is that corporations have vastly more power than individual citizens or even communities in the form of both money and political influence. That decision was inherently anti-democratic, and it led to significant corruption of political authority.

The Citizens United decision shifted social control from partially serving the interests of we the people to corporations, which have overwhelming financial power to achieve political ends often at odds with the public good. Power begets power, and without a balance of power, society becomes lopsided, its hierarchy distorted by extremity—and thereby highly unjust. As this trend continues, American society has, itself, become more unstable.

In modern social systems, control increasingly emanates from the top down in a hierarchy of domination that can only claim to be democratic by maintaining the appearance of formal democratic institutions. Sheldon Wolin (2008) calls it Democracy, Inc., that is, a merging of the institutional power of corporations and the ostensibly democratic nation-state. I call it the corporate state. In effect, the modern industrial nation is a political hierarchy of economic power in which the national government and the complex of transnational corporations coordinate their actions to, in effect, rule society. They give lip service to democracy, maintaining its formal structures, but limit citizen participation in choosing between political candidates and parties, most of which are almost entirely beholden to the corporate donors and sponsors who control policy by direct influence on legislation.

In 2014, Martin Gilens and Benjamin Page reported their research into the relative influence of elites, interest groups, and average citizens on federal legislation over decades. What they found was shocking, though not entirely surprising. Compared with elites and powerful business interests, average citizen groups had virtually no influence over legislation or policy. That is what I mean when I use the term corporate state; the state is ruled almost entirely by the corporate interests.

Legislators beholden to elites simply ignored various voter preferences, such as gun control, abortion rights, and voting rights. The federal government maintains democratic formalisms, yet severely suppresses actual democratic processes, resulting in few, if any, actual democratic outcomes. Is it any wonder that wild conspiracy theories about the "deep state" proliferate? Countless public opinion polls show that many of the policies that the public favors are exactly the ones that corporate-sponsored politicians fight against on behalf of their donors.

In an actual practicing democracy, power and policy would not have a centralized, giant, hierarchical institutional complex. Policy and practice would distribute democratic decision-making so that outcomes of the political process would reflect the needs of the citizens of each jurisdiction because those citizens would participate in decision-making. In that way, democracy would sustain communities, regions, and the larger society at each level, resulting in a strong system integrated by the effective operations of its parts.

An actual democratic society is a stable system because it meets the needs of its people at all levels, due to robust citizen engagement and participation. However, the most recent developments in modern industrial nations have a very different tenor. When societies concentrate political and economic power in the hands of elites, democratic institutions and processes weaken. Then the prospects for political chaos, and the rise of authoritarian control grow.

The weakening of democratic practices contributed to the rise of hate crimes, White nationalism, and the seditious insurrection of January 6, 2021, at the American capitol, led by a president who refused to admit his electoral defeat. In anticipation of losing the election, he had spread the big lie of voter fraud even prior to the beginning of the electoral campaign, and he even attempted to stop the official counting of electoral votes in Congress. The growing chaos and resentment continue unresolved.

As historian Timothy Snyder wisely points out in his small, powerful book *Tyranny: Lessons from the Twentieth Century* (2017), we have plenty of examples of how autocratic attempts (Gessen 2020) to overthrow democracy can succeed until they collapse over their own corruption and hubris. It would be wise to heed Snyder's lessons. Strong citizen action in opposition to demagoguery and autocratic attempts can stop the decline and fall of democracy before tyranny takes full hold.

Stable Systems, Disruptions, and Collapse

Geologists have described the last eleven thousand years or so, which they call the Holocene Epoch, as characterized by its very stable ecosystems and consistently stable climate patterns. Human habitats remained in exceptional ecological balance over thousands of years. As a result, human groups could rely on stable supplies of food sources in those habitats. That enabled all sorts of developments in the human species that amplified our ability to control the factors that determined our survival and allowed Homo sapiens to expand across the planet.

It all centered on the predictability of the environmental conditions humans experienced. That predictability allowed humans to evolve both genetically and culturally in powerful ways. We became more productive in gathering, processing, and utilizing food and materials and making and using tools to ensure higher rates of survival and prosperity. Our populations grew and spread.

Advances in biological anthropology, archeology, geology, and even climate science have assembled diverse forms of evidence, showing how humans came to dominate the habitats in which they lived. The eventual result was a rapidly growing industrial system that now literally consumes the world, not unlike an aggressive cancer, which grows uncontrolled until it kills the very body that sustained it.

The whole Earth System is the body that sustains humanity and can continue to do so if we do not destroy it by overshooting its capacity to carry the load of our extraction-production-consumption waste. Remarkably, the Industrial Age began slowly with the emergence of science and technology and then accelerated rapidly, taking up only about the last two or three hundred of those eleven thousand years of Holocene stability. Given how rapid that growth has been, especially in the last fifty years, it is not surprising that it has gotten so far ahead of us.

However, its greatest impact on the planet has occurred in only the last few decades. The acceleration of economic growth in that short time has decimated ecosystems around the world and caused growing climate instability. As the industrial-consumer economies of the Global North reached the peak of their rational-legal complexity, their institutions achieved maximum control over the societies they dominated. That resulted in growing imbalances and instabilities, which threaten the stability of whole societies and the Earth-System components that the global economy has plundered. As the father of system dynamics, Jay Forrester (1968), pointed out over a half-century ago, the larger a system grows, the more vulnerable it becomes to the danger of destabilization and collapse. The evidence is overwhelming. Both diverse living Earth systems and the global political economy are well on their way to catastrophic failure.

Exponential population growth followed the accelerated advancements in technology and economic organization, which allowed abundance to challenge scarcity. However, by overshooting the Earth's capacity to sustain them, these processes of growth eventually brought on a dangerous New Great Transformation of the complex Earth System itself. Some geologists now call our new epoch the Anthropocene because humans are fundamentally changing key elements of the planetary system more and more each day.

We have begun to experience this new global change in diverse ways around the world. Most of the world's economic, political, and cultural leaders have not noticed or just refuse to believe that we are now already deep into a transformation of our relations with the Earth System, a transformation that we caused but neither understand nor have tried to control. The evidence clearly contradicts their ignorance and denial of the limits to growth. Along with early warnings of superstorms, draughts, and floods, the record-breaking heat domes around the world in the summer of 2023 were only the beginning.

They caused scientists to worry over their interaction with a new El Nino effect in the eastern Pacific Ocean. Repeated atmospheric rivers of vapor laden air have devastated parts of California, where they have unloaded massive amounts of rain. Yet, every word about economics by politicians signifies a commitment to economic growth as equivalent to progress.

The relatively new scientific discipline called Earth-System science (Lenton 2016) examines the entire planet with the goal of understanding how all the subsystems interact with each other to produce the diverse conditions and changes we experience. I capitalize the term Earth System because I want to emphasize its status as a formal entity. It is a complex adaptive living system, and we are a part of it. We capitalize the names of nations, species, tribes, families, and persons. Why should we not capitalize the name of the greatest living system of all, the Earth System?

Systems thinking is a distinctly different perspective than was common in science until very recently. Today, more and more scientists are shifting from a linear paradigm that assumes we can understand everything by tracking causal relationships between individual variables in the world. The systems perspective in science recognizes that the world is far too complex for mere variable analysis to be adequate. Complex systems are far more than the sum of their parts; all their parts interact with one another. To grasp fully the Nature of a complex adaptive system we must utilize nonlinear analytic tools as part of a new paradigm of complex systems thinking and research (Meadows 2008).

Another important term describes the geological epoch we are apparently entering as the Holocene fades into chaos. The name Anthropocene refers to the fact that humans now have the dominant impact on the important components of the Earth System. Some dispute remains as to if we are still in the Holocene or are fully within the Anthropocene. However, for the purposes of seeking to understand the impact of humans on the Earth System and the effects of those changes upon human life, we have definitely entered the Anthropocene Epoch.

Geologists debate exactly when the Anthropocene Epoch began. That is a subtle and difficult problem to solve technically. However, it is far more important that we recognize what we must do now about our situation and its current direction in relation to our planetary habitat. Humans have influenced important elements of the planet for quite a while. However, it is now abundantly clear that the bulk of that influence has occurred during the Industrial Age, most in the last half-century or so and most profoundly in just the past few decades. The changes we have wrought are far overreaching our willingness so far to respond to the dangers they entail.

That influence is accelerating exponentially, even as some of us begin to recognize it and talk about the importance of restraining it. Influence is one thing; transformation is quite another. We are living through the early stages of a New Great Transformation of the Earth System itself. This will likely have a vastly greater impact on humanity than the societal transformation caused by the Industrial Revolution—and we are woefully oblivious and unprepared.

The direction of this transformation of society in response to the destabilization of the Earth System will depend on how, and how quickly, humanity responds to the existential threat posed by the extreme changes in biosphere conditions, which ultimately determine our fate. I cannot imagine anything more disruptive to modern human life than the tragic transformation of the Earth System, as we have already begun to experience it.

If we think we can ride it out or simply fail to decide, we will certainly face societal collapse, even possible species extinction, as extreme conditions become unlivable. And, oh, by the way, indecision *is* a decision in light of the probable consequences. Or we could decide to transform human institutions radically, as top climate scientists (Steffens et al. 2019) have advised, to be able to both mitigate the dangerous conditions we have caused and adapt to the conditions that have already become almost impossible to mitigate.

The rational choice would be to take the extreme measures now necessary. Yet, they will be incredibly difficult to execute. The more we fail to mitigate, the harder it will be to adapt to ever more extreme conditions. The changes we need will be extremely difficult. They require that we downsize many economic processes and transform the energy inputs of others (not just by converting to clean energy production but by conserving a very large portion of the energy that we now waste on unimportant products and services). This is where we must make some very difficult decisions.

This always makes me think of the amazing collective conversion that industry underwent when the US entered World War II. Many other factors existed, but the danger then was immediate and so obvious to everyone that collective, national action became imperative for almost everyone. The danger now is equally immediate, and the threat vastly more existentially grave, yet, because of the complexity and forces of resistance, not nearly so obvious to significant numbers of people.

The comfortable conditions of the Holocene are fast disappearing. The nature and consequences of the modern industrial societies—and the other societies that have contributed very little to the destabilization of the Earth System—will depend largely on the human response within the industrialized nations. Will the industrial nations transform themselves enough to slow the degradation of the Earth System to a point where the Earth can remain a habitable place that we can call our own? That is the epochal question of the twenty-first century. It is a matter of where the locus of social control lies, now and in the near future, with the people and new creativity or with the moribund, predatory, extractive institutions of the past.

Our planet is our habitat, and we depend on its stability for our health and survival, yet we are severely damaging our habitat, and the consequences of the environmental damage from the Industrial Era are rapidly becoming ever more severe. So far, our response remains either indifferent, contemplative, or inactive, except for a few who object, such as members of the Extinction Rebellion, the Sunrise Movement, and Greta Thunberg's Fridays for Future school strike for climate action. These protest movements are great for drawing attention to the Earth-System crisis, but they only point to needed change. They cannot make the actual change we need on their own. Even so, we need much more of them.

As evidenced by major disruptions of climate and ecosystems, terrestrial and oceanic and atmospheric, the very Earth System itself is already undergoing its own transformation from system-stability to destabilization. The results of Earth-System destabilization are hard to predict in exact detail, but the general trend does not bode well for human survival. As the saying goes, "If you break it, you buy it." The Earth System is priceless. Most of humanity has yet to acknowledge our severe disruption of the Earth System and our responsibility to restore that stability to a point where humans and many other species can survive on a livable planet.

In many ways, the Industrial Age has dramatically changed conditions in each of the major subsystems of planet Earth, transforming many of its elements and their relations. One of the earliest signs was the disruption of atmospheric conditions and the extreme weather events we now experience, caused by the effluents of industrial modernity.

Our elites' intentions lay elsewhere: economic growth for capital accumulation and the proliferation of consumer goods for profit. Both resulted in the concentration of wealth among a small elite population and, ultimately, the shrinkage of the middle class that had blossomed in the post-World War II era, at least in the US and Europe. It all presumed unlimited control over Nature and, therefore, control of our future. Growth, the elites keep claiming even now, will resolve the extreme disparities of income and wealth, yet the "trickle down" theory of economic distribution promoted by the rich has never worked and never will.

That is the part where we blew it big time. As nations, we still have little understanding of the changes we have created or those to come. So-called leaders cannot accept that our hierarchical societies have caused them or that we are not really in control of the changes we have already caused. Nevertheless, these changes will involve a great deal more instability, which we have hardly begun to consider, no less mitigate. Significant population decline, for all the wrong reasons, is almost a certain consequence of the growing disruption of the biosphere, as well as the severe food insecurity and social conflict it will cause.

Even long before the Industrial Revolution, the emerging practices and knowledge of Western science and culture had abstracted humanity from Nature. Humans, particularly in what is now called Europe, developed cultural precepts closely tied to a hierarchical form of Christianity that defined humans as distinct, separate from, and in charge of the natural world. That is where much of the trouble began. The subsequent brief period of material success that is now floundering only drew the members of industrial civilization further apart from Nature in both their minds and their practices.

Science and technology driven by finance capital took greater and greater control over the world throughout the Industrial Era, but especially in the past fifty years. Their material success created an endless growth machine within the confines of a finite planet that could not sustain it for long. Yet, the absurd irony of presumed endless growth on a finite planet still escapes most mainstream economists. Eventually, disturbances and perturbations in natural systems began to challenge the stability of both.

Perhaps our biggest problem at this point is that we built our belief systems around two flawed assumptions. The first flawed assumption is that a stable environment with an unlimited supply of energy and materials even exists or could continue indefinitely. The second foolish assumption is that our entire political economy can stand on the belief that our finite planetary habitat, which we somehow see as separate, can endlessly tolerate our accelerating exploitation of it.

More and more facts and environmental catastrophes expose the absurdity of both assumptions as we encounter the planetary limits to economic and population growth, as well as the increasing probability of societal collapse. Yet, that absurdity is far from universally accepted.

Neither assumption bodes well for societal stability, the weakening of which we can see in many sectors today. The question of whether or not we can hold it together under increasingly extreme conditions and make the right choices remains unanswered—except by pessimists and optimists whose fatalism assures us of either doom or utopia. Both live in denial of reality. The answer depends entirely on if we can make the major changes necessary now.

Elements of Stability and Change

Change is everywhere all the time; so is stability. That may seem to be a paradox, but it is not. Just look up at the sky where all the stars seem static unless you wait or look around you right where you stand. You can probably see something moving if only the tree leaves fluttering in the breeze. Some change is very slow, some explosive. You can't see the tree growing on any given day. But what about that truck careening around the corner toward you as you stand in the crosswalk?

Change and stability fluctuate over time and under diverse conditions. Much bigger changes can be hard to see except over extensive time. But they may be just as deadly. That is how it is with climate change, previously measured over thousands of years, until lately. On the other hand, social change can be fast or slow. Some fashions change overnight, and some revolutions occur within hours. Some social changes encounter great resistance and slow greatly or stop.

With social change and social order, things are very different now from when I was growing up in the 1950s. I didn't really notice much change around me until I moved from California to Texas, then Florida and Montana, and as I hitchhiked through several other states while in the Air Force. Those changes had to do with location more than time—of course, at that age, I had not yet experienced much of time or place. Much later, I could not miss seeing some of the changes that were occurring in the South during the civil rights movement and the social conflict over the Vietnam War. As my observational skills developed further, some social changes became obvious that had not been before.

Today, we have entered an era where the changes to our planet are becoming globally severe, even though we could not see much of it initially without careful scientific research. That is only part of the reason why "global warming," now called "climate change," was subject to so much skepticism at first. A more accurate term might have been "overheating the planet." Later, it became obvious that much of the resistance to acknowledging these global changes stemmed from propaganda, fear, and powerful economic interests fighting to retain the status quo. Admitting to these global changes meant recognizing that we will have to change the way we live in very drastic ways. That is not easy for most people to accept, especially when they have so much invested in the world as it was.

However, after decades of collective indifference, we are now experiencing a complex convergence of several forms of change in multiple components of the Earth System as we once knew it. At the same time, societal chaos seems to escalate. As the effects of climate chaos and ecological damage become more and more obvious, it is harder and harder to ignore the larger scale of change. Intense firestorms in California and the Southwest were previously unheard of. So-called "one-hundred-year floods" occur every few years. The list goes on …

A while back, I heard of atmospheric rivers. Weather reporters use it to describe the massive transport of water vapor from warming seas by high winds that roughly parallel the jet stream from west to east. The jet stream itself oscillates in its path in unusual ways, sometimes resulting in a polar vortex, dragging arctic air as far south as the shores of the Gulf of Mexico.

In 2023 and again in 2024, atmospheric rivers dumped more water onto California than anyone could remember. It is a direct consequence of the heating of the atmosphere and oceans caused by carbon emissions. The amount of water the air can absorb grows exponentially with every degree of temperature increase.

Climate scientists had decades ago predicted increasingly erratic weather patterns caused by heating the atmosphere. Well, here they are, so buckle up. If there was ever a concept from mathematics that everyone should understand today, it is exponential change.

Depending on their age, most Americans today are aware of significant changes, such as the end of World War II, the Korean War, the Cold War, the rise of the counterculture and the civil rights movement of the 1960s. They also acknowledge the microelectronic revolution that has dominated consumer culture since the 1980s. In a way, it is all relative to your personal lifespan. Mine is now rather long. I actually remember, as a toddler, seeing people running into the street in our Southern California neighborhood, cheering the end of World War II in 1945. Of course, I had no idea what all the ruckus was about, but the level of ecstatic joy made it a permanent image in my memory.

Most of the world-shaking events of the past century or so were singular occurrences or a brief period of a few years when a cultural change or armed conflict had a relatively clear beginning and end. Today, it seems that everything is changing, stability is lost, and there is no end in sight, just more rapid change. For most folks, that is a very scary realization, regardless of how well they understand what is happening. Yet, here we are.

Social Control under Duress

There are many kinds of stress. Some are good for us, others not at all. I remember reading a book many years ago by Hans Selye, *The Stress of Life* ([1956] 1978). Selye pointed out that stress is an integral part of life but that modern life causes excessive stress over long periods of time, which can be very damaging. Extreme sports junkies love the adrenaline rush of skiing down a steep slope with treacherous leaps over cliffs—as do base jumpers who fly off a ledge in their flying squirrel suits skimming along steep cliffs before pulling their parachute ripcord just in time to land in the valley below.

I have always found the challenge of landing a small airplane in a strong crosswind, reaching for the edge of my skills, to be quite exhilarating. Pilots joke that a good landing is any landing after which you walk away. Such activities are very intense and very short-lived, allowing us to match a challenge with our skills. They do not go on for more than a few minutes, after which one can reflect on the experience and move on. That is the kind of stress our ancestors experienced. Experts say that getting into the flow involves reaching for the edge of your skills and doing something challenging that you really enjoy. The exhilaration of that kind of stress can be a very positive experience. Not so with chronic stress.

The stress of sitting in an office all day, knowing that someone, or worse, an AI algorithm, monitors your every move, is very different. The same idea applies to working on the shop floor in a factory, especially if you are under constant surveillance, such as in an Amazon fulfillment center or perhaps a Tesla factory. The stress is continuous and quite the opposite of exhilarating. Constant unresolved stress damages mind as well as body.

Then, of course, there is the constant tension of having to work three minimum-wage jobs just to cover the rent and not enough cheap, not-so-nutritious food. Such ongoing experiences build continuous tension in your body, which is hard to disperse. It stays with you unless you take some tension-releasing action soon. Even the stress of working as a high-end lawyer earning lots of money takes its toll. Often, these diverse forms of chronic stress led to "self-medication" with illicit drugs or alcohol.

Not long ago, a truck driver delivered two pallets of bricks for some patio work I was doing in my backyard. He was clearly in a rush and began to drop the first pallet onto the driveway before I even knew he was there. When I saw him there, I rushed out to tell him that I needed the materials at the side of the house near the gate to the backyard. Otherwise, I would have a lot of bricks to hand-move a lot farther. He seemed very annoyed but did as I asked.

Later, once he had deposited the materials, I thanked him for taking the time to put the materials where I needed them. He explained that his frustration was because his supervisors tracked his every move via GPS. The company allotted him only so much time for each delivery from his large truck, based on the distance driven between each delivery, or they would dock his pay. His stress was a constant consequence of an economic system driven by capital accumulation rather than human wellbeing.

When I was teaching at the university, after a hard day dealing with frequently unprepared and sometimes arrogant students, along with the occasional ignorant autocratic administrator, I was so happy to go to the dojo and practice Aikido for three hours. I would begin the practice feeling the tension of the day in my body, then practice for hours with my friends and come out tired, relaxed, energized, and relatively elated. The tumbling and break-falls were especially helpful; they literally shook and pounded out the stress of the day.

The point of all this is to say that modern industrial-consumer culture does not offer as happy a condition as the advertising industry would have you believe. No matter how many products you buy, they will not rid you of the chronic stress built into the industrial-consumer "lifestyle." Neolithic hunters and gatherers enjoyed plenty of leisure activities, which if observations of their more recent generations indicate, they enjoyed with a calm, peaceful outlook. The cheerful demeanor of the !Kung, isolated from modernity until after the 1950s, despite the difficult conditions for survival, paints a very different picture of not only social control but contentment with life. This contrast is not so often seen on our freeways or in our offices, factories, or job sites.

Almost everyone today complains about the stress of everyday modern life; increasing numbers suffer both economic degradation and social denigration. Yet, others revel in the adrenalin rush of their high-stress, high-income positions until that heart attack. The consequences of what we have lost in our quest for "progress" are profound. It is not a stretch to say that industrial modernity is out of control.

When I heard about the Blue Zones several years ago, I had to learn more. The Blue Zones are places on the planet where people are the happiest and tend to live much longer than elsewhere. How could we not ask why? As it turns out, in the Blue Zones, many more people live past one hundred years than anywhere else on the planet.

Dan Buettner (2012) traveled the world for National Geographic to research the happiest people on Earth. What he found had nothing to do with the progress of industrial consumerism. In subsequent years, he and others have added much more information about happiness, health, and longevity at the website: *https://www.bluezones.com/*

In the Blue Zones, Buettner found that both happiness and longevity have a lot to do with community and lifeways. (I don't like the more commonly used term "lifestyle," which seems to hint at fashion, not ways of living and relationships.) People live longer and healthier because of a few simple habits and by fully engaging with others in the community where they live. Researchers discovered that the Blue Zones are in several very different cultures around the world, from Costa Rica to Okinawa and from Sardinia to Loma Linda in Southern California. It is not so much where they live as how they live in relation to their communities and the world around them.

Think about all this in terms of the locus of social control. People who live in the Blue Zones are clearly in direct personal control of their own lives and are able to make choices that enhance their health, happiness, and longevity with the support of family and community. Compare America's sacrifice zones, where residents are not only isolated from economic opportunity but also deprived of the very basics of a healthy life. The differences are profound. Even in the suburbs, people often do not even know their neighbors; they have no real community. They have high mortgage payments and a McMansion to sleep in before hitting the freeway again.

The dominant ideology of the American hierarchy is the "trickle-down" theory of economic progress. We know that a key characteristic of hierarchy is that over time, power and resources tend to concentrate at the top. In the early decades of the twenty-first century, as the concentration of wealth, income, and political power surpassed that just before the Great Depression of the 1930s, new tax breaks, loopholes, and subsidies for the largest corporations and the super-rich reached an all-time high. A sudden windfall profit does not relieve chronic stress. At the same time, through automation and the outsourcing of labor to the cheapest global labor markets, the American middle class has all but disappeared, contributing to chronic stress and the ill health that usually entails.

The idea that giving all sorts of special privileges to the wealthiest persons and corporations in the nation will somehow trickle down to the most isolated and vulnerable populations in the land doesn't even pass the preliminary smell test. Moreover, decades of evidence confirm the obvious absurdity. The rich hold on to their money with great enthusiasm. Often quoted, Lord Acton stated, "Power corrupts, and absolute power corrupts absolutely." In fact, power is usually used to gain more power, and lack of power causes the powerless to have even less power over their own lives over time. Money is power.

Social control now concentrates extremely at the very top of the increasingly anti-democratic hierarchy of an ostensibly democratic society. Democratic forms are retained, almost as a matter of style, yet corrupted so severely that democracy is hardly recognizable anymore except as an empty ritual. We are long overdue for a complete redo of the American political economy.

CHAPTER SIX

What Economies Are For

ANY LIVING SYSTEM that evolved in the context of a relatively stable environment is likely to remain stable as long as its environmental conditions and habitat also remain stable. That is simply because of their co-evolution. Living systems can also evolve effectively when the conditions of their habitat evolve if they co-evolve slowly.

When human groups were small and co-evolved slowly with other creatures in their immediate ecosystem, stability reigned for eons of geological time. Change was slow and incremental. Biological and social evolution occurred together. However, once increased control over their environments allowed societies to grow larger and venture out into new territories, their situations became much more dynamic. The pace of change increased, and human groups had to adapt more rapidly, even though change back then might seem incredibly slow to us moderns.

Our modern industrial culture leads us to believe that there are economic laws that are universal and inviolable. Well, the economics profession probably had as much to do with shaping those economic assumptions as anything. When you think about it, mainstream economic theory has groomed politicians to accept assumptions about the relations of business to society as a whole, which clearly favors the ascension to increased power and status by financial and commercial elites. Those in power were not about to dispute such assumptions. Today, economics is the most powerful academic discipline among the social sciences. Why? Because it cultivates financial support from the elites, who benefit from the application of conventional economic theory to politics in favor of the rich and powerful.

Now, don't get me wrong. Some very fine work has come out of economics, especially considering the intellectual-political constraints on the profession. That work is important for our discussion in this chapter regarding the role economies play in politics. However, the consensus among mainstream economists derives from a long-standing ideology that has little empirical evidence to support it, and it is leading us down a very dangerous path.

The model of classical economics (read Adam Smith and his invisible hand, among others) and modern neoliberal economic theory (Milton Friedman and the Chicago Boys) both rely on the same myth. We can excuse Adam Smith for this because the world about which he theorized fit his model rather well at the scale of a small community when he published *The Wealth of Nations* in 1776. However, it does not fit our situation two and a half centuries after his book and the American Revolution.

In local communities, like a village or town with its surrounding agrarian population, where each merchant, tradesman, and farmer serves their own interests by producing useful products that benefiteveryone through economic exchange, Smith's theory worked well in the pre-industrial context. However, Smith also recognized the role of moral sentiments in human behavior and in social relations. He also recognized the dangers of the growing power of large corporations in his day, which those who tout clichés about his ideas do not. Neoliberal economists today ignore those two concerns of Smith.

Today's mainstream economic pundits love to cite Adam Smith as applicable today despite the fact that doing so takes his analysis entirely out of context. In Adam Smith's day, the larger culture embedded economic activity within the societies in which it occurred, although that was beginning to change with growing international trade. Strong moral precepts framed most economic behavior, which is not to say that no one cheated or priced-gouged if given a chance. But in a small society, social control is cultural and interpersonal. It involves values and rules of mutual respect, accommodation, and moral principles that cannot be overridden easily. That has all changed.

The "rationalization of society" and the primacy of economic power came as essential features of the Industrial Revolution. The rational-legal requirements of economic growth forced society and culture to take a back seat to elite opportunities for capital accumulation beyond colonialism, which were the driving force behind industrial-ization and related international trade. You might say that social control shifted from traditional culture to economic interests.

The need of the wealthy and powerful for capital investments increasingly controlled society, while traditional cultural values that supported the needs of ordinary people controlled less and less of the decisions that affected everyone. Given the power of that fundamental shift, it is remarkable how many traditional, some would say univer-sal, cultural values still survive among the people within industrial societies today, even as those values are beset with confusion in modern culture.

It is common today for businesspeople whose sole interest is sales, profits, and expansion to try to associate their economic interests with traditional values and "personal freedom," "family values," and other post-enlightenment ideals, which they try to identify with their own economic interests. These ideological efforts align closely with the "neoliberal" economic theories of academic economists, who promote them as "scientific" justification for the priority of the interests of the economically powerful over the broader interests of the public—

invoking versions of the counterfactual "trickle down" theory of the distribution of wealth.

The rise of multiple intersecting crises of climate, ecology, and survival for people across the planet exacerbates this fundamental contradiction between the dominant economic ideology and the social good. In this chapter, I want to explore the alternatives to the dead-end of the neoliberal economics of endless economic growth for capital accumulation by the global financial elites. Economic policies do not reflect a bunch of universal laws; they are social-political choices made by the powerful.

Those with the most political power choose our economic policies in their own interests. Not surprisingly, that is not working out so well for the rest of us. The locus of social control of public policy must change to a choice made by we the people, for the benefit of all, including all living systems on the planet. That is what economies *should* be for.

From Linear Economics to Ecological Society

The design and operation of modern mainstream economics are a function of the needs of industrial and financial capital. As such, it can, but often does not, involve benefits to the societies in which it operates. If we take the principle of seeking human wellbeing as our core value on which to base our judgments of human action and the systems we create, mainstream economics is a big failure. It remains dominant only because it protects the political economy's hierarchy of power so well.

The ideology of the economic elite claims that "free markets" naturally and universally lead to human progress. Unfortunately, these allegedly "free markets" are only free for the corporations and financial institutions that control them—to the extent that governments do not regulate them in the public interest. But markets are certainly not free for the rest of us. And, although they do, in fact, lead to all sorts of things, few of those things are progressive. Of course, the economic elite deems any government control of corporations to be an evil stifling of *human* freedom—however, it is the freedom of institutional elites and the super-wealthy to control the economy that they seek.

If markets were really free, the financial and banking organizations that caused the Great Recession of 2008 would have collapsed. However, under pressure to keep the economy from collapsing without "interfering" with the economic hierarchy, the political authorities judged these institutions "too big to fail." The federal government bailed them out with billions of dollars issued by the Treasury, adding all those liabilities to the national debt—making everyone responsible for the misdeeds of the financial elite. Even more social control than before shifted from the government to the financial elite. The domination of the hierarchy of power by the corporate financial institutions became stronger, and the linear "progress" of economic growth continued. Hundreds of thousands of families lost their homes for that "economic progress." And, despite all the financial crimes, nobody went to jail.

If we look at the trajectory of modern economies since the Industrial Revolution, we can see that just a few elements have determined where we find ourselves now and why we are in such a precarious position. First, the key premise underlying the global corporate economy is that it must grow at all costs because its basis is debt, payment of interest, and capital accumulation.

Capital invested becomes a debt owed by the persons, projects, or organizations that borrowed and invested the capital, plus interest. Because the debtor must pay off that debt with interest, the amount then owed is always more than the original debt. The only way to pay it all back to the lender is to make a profit. Where does profit come from? And what about losses?

The billions of dollars the federal government simply gave to bail out the speculative banks and investment houses in 2008 were created by simply posting the debt to the people's account with the central bank. At this point, the best thing I can say about that and the impunity of the too-big-to-fail swindlers is, "Don't get me started." Anyway, in non-crisis times, the growth of the enterprise pays the expanded debt, including interest. The system requires continued growth to keep operating in this way.

Simply put, the money to pay the debt plus interest and profit to investors comes from economic growth. Expanded production and consumption must generate revenue that exceeds the value of the investment by at least the amount of interest accrued. The returns on investment must be greater than the amount invested or the investor loses capital. Only by expanding production and sales the revenue that results from its growth can capital growth continue. Without growth, there can be no complete payment of the debt plus interest owed. That is why the system of industrial capital requires endless growth. However, this notion is ultimately fruitless on our finite planet. It is a terminal enterprise.

Underlying all this is the fact that we have a debt-based monetary system in which the central bank issues money by fiat, having entered a debit onto the national balance sheet, another way the rich get richer and the nation becomes a pauper. That is another very long discussion too long to have here. Suffice it to say, any sovereign nation need not base its currency on debt to a privately held central bank: the Federal Reserve banking system in our case. It can simply issue currency to whatever extent its economy needs a greater supply of money in circulation, and it can direct that money to the public interest. Instead, the private bankers were able to set up the federal monetary system for their own profit, ignoring the public interest.

The system worked well in terms of expanding economic wealth and consumption as long as the system had room to grow and people tolerated serious income and wealth disparities. However, the most recent inequitable distribution of the benefits of capitalism has become so extreme that the likelihood of broader societal instability grows by the day. Moreover, the ever-expanding corporate political economy has reached the planetary limits to its growth.

Yes, growth can continue for a short while and only by destroying the living systems upon which we all depend. That is the great predicament of our mainstream (neoliberal) global economy. Its premises are false and its linear trajectory is unsustainable once it reaches the planetary limits to growth—which is right about now.

Enter the combined climate and ecological emergency, as if the planetary limits to growth were not daunting enough. Humanity has now reached a turning point, a point of no return, perhaps even a tipping point beyond which recovery of any semblance of stability and prosperity is a seriously open question. The one certainty is that we cannot go on following the path the global political-economic elites claim is necessary and inevitable.

Fortunately, a small but growing number of creative economists, whom I call "outlier economists," have been working in recent decades on potential alternative economic models. Some, like Richard Heinberg (2011), have called for the end of growth in response to the conditions that have evolved since Donella Meadows (1972) and her colleagues predicted that the end of growth would begin in the early decades of the twenty-first century. They base these models on more palatable principles, such as human wellbeing and distributive justice, as the basis for a new economy. The widely misunderstood concept of de-growth (Raworth 2017; Hickel 2021) is a key idea among this new breed of economists and critics of the current dead-end economic theories. Read on.

De-Growth and Regenerative Economics

All living systems rely on external sources of resources and energy to maintain themselves. All the living systems that are subsystems within the Earth System rely on energy from the sun in one form or another. Recognizing that the Earth System and its subsystems constitute complex adaptive systems operating in multiple interrelations makes it clear that balance is central to the stability and continuation of the Earth System and all its creatures.

However, our current trajectory of unsustainable endless growth moves in direct opposition to the Nature of living systems. The only alternative is an economy not based on growth as such but instead one organized to serve the basic needs of humans and other species. A viable economy must recognize the human place in the biosphere; we must design it to sustain human wellbeing and stable relations with others in each ecological niche (habitat) in which we all live.

A number of very smart people, especially Kate Raworth (2017) and Jason Hickel (2021), have written books and articles explaining the necessity of de-growth. Some give lectures on these principles, offering us a rich trove of ideas from which to draw as we face the necessity of transforming human societies to live in balance with our habitat, the Earth System. Johan Rockström, Will Steffen, and others developed a deep understanding of the anthropogenic pressures placed on the Earth System that have overshot the planetary boundaries, causing abrupt global environmental changes (Rockström, Steffen, et al. 2009).

Human activities must not overshoot these boundaries if we are to avoid sliding into trajectories that take us across tipping points beyond which irreversible ecosystem collapse follows. Maintaining a stable biosphere will require collective human action to achieve "... decarbonization of the global economy, enhancement of biosphere carbon sinks, behavioral changes, technological innovations, new governance arrangements, and transformed social values." (Steffen, Rockström, et al. 2018: 8252) A big problem is that we cannot predict the exact sequence and timing of reaching the dangerous tipping points toward collapse. But we do know that we are getting very close and may have already passed some tipping points.

What does this mean? It means very big global changes in both institutions and lifeways, which is why I call it a New Great Transformation—much bigger than the Industrial Revolution. Remember, we are already experiencing a New Great Transformation of the Earth System. We had better own it and take the advice of the scientists who have described it so well, or it will overrun us as the biosphere fully destabilizes, threatening our very survival. We must form regenerative economic systems that can allow the achievement of three key functions. Here are some of the requirements for moving our societies away from our current dead-end model of economic growth.

First, stop the destabilization of the complex Earth System by constraining the industrial-consumer economy. That not only faces a big cultural/behavioral obstacle but also runs into the rigid resistance of the most powerful institutions—and their wealthy owners—in the world.

Second, restore and regenerate as much of the biosphere and its diversity as we can. That will take concerted effort and reorganization of communities everywhere.

Third, build new societal formations that can live in harmony with our Earth System while doing as much as possible to accomplish one and two. That will require a complete cultural reset to replace the industrial-consumer mentality of people everywhere.

Dmitry Orlov (2016) calls the first step shrinking the technosphere. But that will involve many things, the most complex and difficult of which is ubiquitous social change. The disruption of everyday life will be immense, adding to the displacement and food insecurity due to growing climate chaos. Even if we do everything right—and that is a long shot—the human population will sustain significant losses due to the severe disruption of climate and ecosystems already in the pipeline due to past and present damage and destabilization. Our task now is to achieve all three objectives as quickly as possible to minimize suffering while building a livable future. That will require a cultural revolution the likes of which Mao could never have imagined. Those from above cannot decree a real cultural revolution; it must emerge from the needs and necessities of the people or it is not real.

Regardless of these key functions or any other actions humans take, the planet cannot sustain our current levels of extraction, production, consumption, and waste generated by the Global North and emulated by the growing populations of the Global South. Our fossil-fuel-powered technology and distribution systems have increasingly damaged living systems throughout the two- to three-year industrial interlude, with much more intensity in the past few decades. Along with more frequent and intense catastrophic weather events and crop failures, sea rise due to ice melt will force at least partial abandonment of many of the world's greatest cities and river delta settlements in the next few decades. Massive resettlement of such huge populations will involve great stress and conflict and cause significant population declines in the hardest-hit areas of the Global South as droughts, floods, and crop failures proliferate.

For contrarian economists like Robert Heinberg (2011) and growing numbers of ecological economists and climate activists, de-growth is the most important and viable path to climate stability. De-growth advocates such as Hickel, Raworth, and a growing list of others seek human harmony with the diverse ecosystems that our destructive industrial-consumer economy threatens. They would have us shape wellbeing-based economies founded in low energy-use production and consumption.

The contrasting utopian dream of solving all problems with new technology and materials, as well as industrial innovation, continues to dominate mainstream "environmental" thought in the corporate culture. Bill Gates' efforts to organize no less than twenty-eight billionaires to finance the achievement of an "energy breakthrough" represent one of the most dangerous and egregious of this mindset, if only because of their wealth, power, and self-serving influence on governments. A long list of big corporate promoters supports Bill Gates' (2019) self-congratulatory approach to climate change.

These powerful economic interests resist ceding control of the global economy. We must overcome their dominance despite the fact that they hold the most economic and political power. Not only are they a great distraction from following a viable path to a human future, but they are also a direct threat to garnering the material and cultural resources to make that effort work.

Gates' vision for a techno-financial strategy, elaborated in his book *How to Avoid a Climate Disaster,* claims that to solve the climate crisis, we must all invest in new high-tech (high-energy) technologies for clean, efficient energy production to meet projected demand. A big part of the underlying problem, of course, is growing excessive demand—a boon to financial elites and disaster for society. Gates would compensate for the failure to conserve energy by developing expensive, unproven new energy-intense carbon capture and seques-tration technologies. His naïve projections of energy demand based on current trends in the industrial-consumer economy will fail as the system destabilizes as it further overshoots the fully verified planetary limits to growth (Meadows et al., 1972; Rockström, Steffen, et al., 2009, 2018; Merz, Barnard, et al. 2023).

Instead, a truly civic-minded Gates would have focused his attention and resources on conserving energy, consuming less, deploying low-carbon technologies, curtailing waste, and redistributing and redeploying the elite-controlled wealth of the nation. However, reducing demand for unnecessary energy consumption never occurred to Gates, a growth addict. He and his class of high finance Davos Men see no new profits for themselves in that. The public good is only a slogan for them, framed in their worldview of techno-financial growth seen as the only path to "progress." Of course, Gates' investments in the technologies he promotes would produce handsome profits from the policies he advocates and produce large additional carbon emissions and other pollution associated with industrial developments.

Several foolish assumptions of the economics-as-outside-of-nature illusion led to disaster. One is the belief that we can overcome any problem with a yet-to-be-invented technology or industrially applicable yet-to-be-discovered new material. Neither a sci-fi escape to a new planet nor another fantasy of overcoming Nature with new technology can change our interdependence with the living Earth System or our dependence on finite planetary resources.

The world does not need more high-tech energy production projects. It needs low-carbon retooling to generate energy for the tasks that match the requirements of restoring ecosystems and human habitats. None of this requires much further research and development except for the extremely important task of developing the *societal* strategies to implement human-scale ecologically sound economies.

The denial of Nature is one of the most flawed, though widely held and mostly unconscious assumptions, underlying mainstream economics and public policy. The debt-based monetary system of modern growth economies is ultimately incompatible with the complex biophysical systems of the planet. Simply put, the physical and biological systems in which we live have limits. Those limits are not negotiable; they are the existential context of human life. In their abstract formulations, monetary policy and financial growth have no limits. Mother Nature does.

The unbounded financial manipulation of modern economies by Wall Street's self-anointed "masters of the universe" allows and enables unbounded merciless ecosystem destruction around the world. Over a decade ago, Philip B. Smith and Manfred Max-Neef (2011), two prescient outlier economists, offered one of the most powerful analyses of the core illusions of mainstream economics and its socially destructive consequences. They exposed the inner flaws of the power and greed of neoclassical economics, much like E.F. Schumacher (1973) had done decades before. They argue that compassion and the common good must become the value-bases for an economy that serves the interests of humanity rather than the goal of accumulating wealth and power in the hands of a few. That premise points to an entirely different sense of "appropriate technology." Schumacher had pinpointed that core failure of mainstream economics—it has no soul.

Schumacher's vision of Buddhist economics, grounded in human compassion, had the goal of achieving a livable quality of life for all. That offers an extreme contrast to the idolization of personal and organizational greed in opposition to others, which characterizes the corporate economic worldview that dominates industrial-consumer culture today.

To seek any or all technologies to automate production to save labor, leaving workers to perform only a few menial tasks is not an appropriate technology for a low-energy economy premised on Buddhist economics—just the opposite. Instead, appropriate technologies would consist of those techniques, tools, and equipment that enhance the human experience of work and its products by allowing the development and exercise of personal skills, craft, and creativity.

Instead of growing reliance on external energy sources to drive automated equipment managed by just a few technical workers, appropriate technology would engage workers in creative ways. Such ways would inherently involve more labor, not less; it would also call for more learning, creativity, and craft. Economic de-growth goes hand in hand with developing technologies and production processes that enhance human creativity and wellbeing because craft and high purpose would drive production instead of immediate efficiency and short-term profit. Clair Brown (2017) offers a more recent framing of Buddhist economics built upon Schumacher's work and the principles of equality, sustainability, and right living.

The battle lines are drawn. The survival of ecosystems and their inhabitants, including us, on the Living Earth (Gaia), will require the death of neo-liberal economics and the global corporate industrial-consumer political economy it supports. That death is as complicated as it is inevitable and will result from a globally traumatic process of creative destruction—unless it is left to collapse on its own as it further destabilizes the living Earth's ecological systems. Its timing and character call for rapid and comprehensive human intervention and creativity to offset the trauma of destruction.

The very act of collectively creating a wellbeing regenerative economy will not only destroy the dysfunctional and unsustainable neoliberal economy that mainstream economists hold so dear. That process will also determine our chances to restabilize planetary ecosystems and the climate and for human survival beyond sustaining a few scattered enclaves of suffering climate refugees.

The Economics of a Viable Social Ecology

An economy that can provide necessary goods and services while sustaining its habitat will be no simple thing to achieve. We are starting from less than zero. Industrial civilization has already overshot six of the nine limits for planetary stability (Richardson, Steffen, et al. 2023). Without extreme intervention, that will lead to collapse—see Figure 2. The existing mainstream global political economy is both predatory and degenerative.

After all, the economy in which we now live resulted from the colonization and subjugation of the peoples and plunder of materials from most of the world. It continues to subsume societies and resources under its domination. Its primary value and goal is to accumulate capital by expanding its investment in extracting value from every corner of the planet. Its extraction and deployment of fossil fuels serve that purpose while destroying ecosystems everywhere. In contrast, we must create an ecologically regenerative economy that uses near-zero fossil fuels and far less total energy than the current mainstream economy does today.

That will inevitably entail quite a transformation of society itself, founded on very different principles. That is, ecological principles require us to pull back from the current increasing levels of overshoot in order to achieve a stable societal existence grounded in Earth System stability. When our economic activity is no longer driven by high-energy fuels but rather by the combination of new adaptations of modern and traditional technologies with human and animal power producing the bulk of high-quality needed goods—instead of high-turnover low-quality goods—we will organize our lives quite differently in ways more amenable to engendering human happiness.

Two basic principles must underlie an ecological economy that sustains both a viable human population and its ecosystem. First, a new ecologically viable economy must arise from the underlying value of supporting the wellbeing of its people within a healthy biosphere. Second, it must support the ongoing stability and viability of the biosphere itself. Human economic activity must become an integral part of a self-sustaining biosphere. In contrast, the existing global economy is a predatory colonizer of all planetary resources. To achieve a viable social ecology, we must decolonize the world, which means radically changing our relations with the Earth System from predatory to sustaining.

Economics does not operate in a political vacuum. That is why I use the term "political economy" here. In order to achieve a viable ecological economy, politics must radically change just as much as we must change the economy. Ecological viability inextricably interconnects with economic sustainability, and we are all involved. Ecological viability must become a political priority.

An ecologically viable economy is one that takes no more than it gives back to its habitat. To be ecologically viable, an economy must maintain itself as a species-specific subsystem of the whole society and ecosystem in a harmonious relation to the entire Earth System. Economies, as open systems, must exchange material and energy with their habitat. To operate effectively over time, an economy must sustain its exchanges with its habitat in a balanced way. This, interestingly enough, is where indigenous wisdom meets modern science.

Indigenous peoples typically engage with their habitats on an equal footing. Indigeneity (Delgado-P. and Childs 2012), as I understand it, is the condition of being at one with the world in which we live, recognizing the need to harmonize with our habitats. Indigenous cultures consciously maintain a balance with all the living systems that surround us because people and habitats have co-evolved for millennia—their source of viability. Today, humanity is in a quite different condition. Our economies have evolved as dominators in opposition to the living systems they plunder by overshooting the boundaries of our place in the Earth System. That cannot stand.

To be indigenous is to recognize that we are all elements of the web of life in which we participate and to act in support of the ecosystems of which we are a part. A viable social ecology of community implies that the humans involved fully recognize their place in the ecosystem, not as dominators but as compatible members. In that sense, we must all become indigenous now (Christie 2014).

To develop an ecologically sound economy, we must understand the character of the economy we already live within; only then can we transform it. Our forbearers built the modern global economy on the domination of the world through its early colonization. Although the form has changed—first, from military colonization to political administration, then to political-economic subordination—the global political economy still operates by domination and plunder. Modern domination is sometimes subtle and sometimes violent.

To build a viable human social ecology, we must replace today's hierarchy of power with an ecological relation of mutuality. Only then can we sustain right-sized human populations into the future. This will require us to engage in creating a new type of human society to reflect our understanding of the Nature of our existence on planet Earth. The new economy we create must rely on mutual support and the principles of social ecology in harmony with Nature itself, something currently rather foreign to most "moderns."

We must cast off the chains of illusion that proclaim us separate from both Nature and each other. That dualistic linear thinking at the center of modern civilization has driven the trajectory of the Industrial Era toward the abyss of collapse, leaving us no choice but to abandon it completely. We can and must create a new form of economy designed to enhance the wellbeing of us all, including all the complex living systems with which we interact, by recognizing the unity of all beings in the integrated mega-system of life that we call the Earth System.

CHAPTER SEVEN

Cultural Values, Laws, and Force

WHO CONTROLS WHAT and whom is the age-old question of how to coordinate human behavior for the benefit of both the group and each individual. One form or another of authority most often resolves that question. Authority is all about how leaders make decisions in a group, organization, or state, who gets to make them, who benefits or not, and who must carry out decisions. As historical and contemporary examples demonstrate, the form of authority always falls somewhere along the continuum between dictatorship and democracy. Political power is a self-amplifying feedback loop; power facilitates gaining more power. Unchecked, the power of politicians in a democracy can eventually lead to dictatorship.

Of course, it is not quite that simple. We can see all sorts of variations throughout history. A common, almost universal variation involves slanting public decisions to some degree to favor those who have the most power and can influence decision-making in their own interests.

This chapter deals with both the sources of social control under various forms of authority and how societies respond when things go wrong. In a perfect society, there would be no need for law enforcement since consensus on what is right would also be perfect, and behavior would presumably match that consensus. (Now, that *would* be a utopia!)

However, the real world operates rather differently; our everyday lives and the societies in which we are citizens are far messier than that. Today, the degree of societal consensus declines as economic, ecological, and political crises loom larger and larger. The distribution of power becomes more and more hierarchical as the need for real democracy grows more urgent.

Law enforcement in history has acted mostly as the arm of violence used by those in power to assure compliance with decisions emanating from the top of the social hierarchy and to maintain public order. In ostensibly democratic societies, the enforcement of laws implies that such enforcement expresses the will of the people and the society's moral framework, as enacted in legislation. That, too, is an ideal that may or may not reflect political reality in different situations. Social control varies with the ongoing struggle between hierarchy and democracy and between consensus, indifference, corruption, and downright criminality. Considerable differences between street, financial, and political crimes reflect the relative power of the criminal in the political economy.

If we think in terms of systems, we might ask, how does a given system work best? Well, the obvious answer is that a system runs smoothly when all its parts (subsystems) work in harmony with a clear mutually agreed mission. The near-perfect coordination in the performance of the Air Force Thunderbirds results from each member performing his functions in ways that support the continued operation of the whole system and each of its other parts.

Of course, that is a very narrow example of social control in a highly specialized social unit that has one clear-cut mission. Modern societies are vastly more complex, with many more interconnected parts and purposes. Some complex societies coordinate behavior more effectively than others do because of a high level of consensus, so the group performs better under conditions of minimal conflict. Other systems struggle to hold it together at all.

The authoritarian critique of democracy claims that hierarchical authority is more *efficient* at responding to problems—as perceived by the ruler(s)—and achieving important goals such as providing what the autocrat decides society needs. Authoritarian regimes claim to be superior to the messy process of achieving consensus and collective action in a democracy. However, the notion of a benevolent dictator is a fantasy, rarely seen in a brief moment in history. Most autocrats prove themselves wrong in that they exploit rather than enhance society's wellbeing. Democratic systems are far more effective in serving the wellbeing of their members. Fascists are more interested in glorifying their maximum leader.

Autocratic systems operate in the interests of the autocrats, usually in the guise of benevolence. On the other hand, many human groups have operated based on cooperation and mutual aid since long before the dawn of recorded history, as demonstrated by a growing body of evidence from evolutionary archeology and biology.

This raises the question of social evolution, human biology, and the elements of habitat, which would most likely have changed very gradually over many thousands of years. Some fascinating work in evolutionary anthropology suggests that the use of fire and cooking had direct effects on the human jaw and the amount of time available to pursue activities other than to acquire and digest food (Wrangham 2009). Many such activities involved more use of the brain and more socializing around food. Gastronomy may be the glue that builds social cohesion, which helps explain why so many who can afford it pay a great deal of attention to the endless variety of foods, recipes, and restaurants.

Cooking was a liberating force because it performed a significant part of the digestive process, freeing the jaw from so much work. The shortened time it took to digest allowed much more time for other activities that, in turn, affected the size and performance of the brain.

It also became central to the process of social bonding among members of a group.

Of course, human physiology and culture have since moved far beyond those early developments. Today, most of us moderns are several steps distant from the sources of our food, for which we depend mostly on complex institutions of production and distribution. Too many eat on the run, and that has negative health effects.

In the hectic lives of most employees, long meals with extended conversations and the intimate relations they enable are increasingly rare. Eating on the run has become a detriment to the social order, just as the industrial processing of food is a detriment to nutrition and a contributor to disease.

I remember quite clearly my experience living with a Mexican middle-class family in the summer of 1964 while taking language classes at the University of Guadalajara to fulfill the language requirement at the University of California. We all took the main meal of the day, *comida* (dinner *en inglés*), slowly amid extended conversations about many things, beginning in the mid-afternoon and lasting from about two to five p.m. Expectations for attendance and participation were very strong—*comida* constituted a norm of sociability no one disrespected. People made every dish from scratch and savored them slowly during the course of the conversation.

In addition to family members at the long dining room table in this large home in a quiet middle-class neighborhood, diners included a couple of borders in addition to this American student. One was a Mexican Airlines pilot whose intermittent residence depended on his flight schedule; he had a wry sense of humor. Another was a tall young man who worked in a shoe store and carried a revolver under his suit coat. I learned far more conversational Spanish in that diverse setting than I did in all the classes I had taken combined. I also learned by direct experience a culture where social relations still had a certain priority over economic pursuits. It was all about interpersonal engagement, in direct contrast with economic roles, which are always about individual performance. When the young shoe salesman had to leave early to get back to work, it was not looked on favorably.

How did the sociability of group meals come about? Early hunter-gatherer groups had to accumulate food for their sustenance day to day. Fire and cooking brought people together to eat and tell stories. Archaeological and anthropological evidence suggests that hunter-gatherers had far more leisure time than modern worker-consumers do. Gradually, released from spending so much time acquiring and eating food (after the advent of cooking), human cultures began to flourish.

Feasts after a successful hunt were the highlight of leisure, which allowed for extended socialization, sometimes over several days. Gradually, social organizations became more complex, especially once the agricultural revolution resulted in significant surpluses of food and kinds of work that were more specialized. As more complex forms of society emerged, some were more hierarchical, and others had more widely distributed networks of relations and authority.

Among humans, the way a system works predicts happiness well. Depending on the culture, somewhat different systems may produce an equal amount of happiness for their members. Yet, happiness apparently has some universal characteristics. Throughout history, various forms of social networks and hierarchies have contested prominence as the primary medium of social control and even happiness (Ferguson 2018). It's all about the organization and exercise of authority—and about human needs, values, and desires—and how well they are met. Societies can be held together by various means, some more pleasant than others. But happiness prevails only when relations are grounded in trust and respect.

Authority is legitimate power, not power alone. You could look at authority as one of the important kinds of glue that holds society together. Interpersonal relations are a key source of social cohesion. We bond with people we both know and trust, either personally or at a distance.

However, it is more complicated than that. What makes power legitimate? Well, that varies too. Trust and commitment compose social bonds. If people are willing to comply with the decisions and actions of those who make decisions, that compliance implicitly legitimizes the power of those in control.

Nevertheless, compliance does not necessarily mean consent. We may comply even though we do not agree because we do not have the power to refuse or simply because the cost of refusal is too high. Or, as John Lennon put it, "They've got all the weapons; they've got all the money. It's all there, (in an interview by a fourteen-year-old boy, filmed shortly before an anti-fan assassinated Lennon.)

So, the basis for practical legitimation may range from enthusiastic agreement and commitment to compliance in response to force. If compliance is only in response to force, then power, not authority, is at play. A look at Putin's "modern" Russia, for example, reveals an extremely autocratic hierarchy where raw power and violence play a more extensive role in social control than does the legitimacy of any democratic formality. Whatever legitimacy exists results in large part from the effects of propaganda and information control exercised by Putin's inner circle of former Soviet KGB officers in the Kremlin. They suppress independent media.

Even in the "Western democracies," imperfect as they are, government and corporate propaganda and information control play a part in the institutional efforts to maintain legitimacy despite policies and practices that often favor elites over the citizenry. Congressional representatives and senators routinely vote for legislation contrary to the wishes of their constituents but favoring the economic or political interests of their wealthy donors. Then, of course, in some cases, there is outright bribery.

To the extent that force, information control, and propaganda hold societies together, they are very weak and unstable. Coordination and control are most powerful when members internalize social control in the form of mutual values, beliefs, and behavioral norms because they believe in the righteousness of the structure of power under which they live. They behave accordingly, even if it is more hierarchical than democratic.

That can only happen when there is a high level of consensus on values and beliefs among the people and the authorities. Both recognize that those who exercise power over decision-making do so based on their shared values and beliefs as the basis of consent. That is why power is exercised legitimately—it constitutes authority. Sometimes, however, the people may be mistaken.

These are not merely abstract concepts of political science or history; they play out every day in the ways that people live. Social control in any particular society may be weak or strong or somewhere in between. It will always involve some degree of either consensus, or not, and thus conflict. In a few cases, consensus may be so low that it leads to rebellion or insurrection. In all this, we can see vast differences between the ways indigenous peoples live even today and the complexities of life for members of most modern industrialized societies.

Social Integration and Coordination in Groups

The effectiveness of a group results from agreement on values and goals and how to achieve the objectives agreed to by its members. There will always be some ambiguities in a situation and differences of opinion as to what to do and how to do it. Personal relationships are also very important, although technically, they are not part of the "rational-legal" calculus. Trust is an often-overlooked factor that results from experience and engagement with others.

Consider how such variations and complexities may play out differently by comparing two extremely different groups: the !Kung bushmen of the Kalahari Desert in southern Africa and the elite aerobatic squadrons of the US Air Force Thunderbirds or Navy Blue Angels.

The life ways of a small band of !Kung bushmen in the Kalahari Desert of Southern Africa, discussed in Chapter 3, certainly involved a hard life by our standards, hunting and gathering their food day by day. Yet, their cheerfulness and calm were striking. Everyone seemed to know exactly what to do and how to relate to the others in the tribe. The !Kung lived in a world they fully understood. Hunting in the Kalahari bush and performing team aerobatics both require precise, coordinated actions. Yet, they frame each very differently. The precise coordination of !Kung hunters is not predetermined like the choreography of team aerobatics, yet their precision is very important for group survival. The pace of action, of course, is different in the extreme.

Reflecting on the vast differences in the lives of the !Kung people from our own made me question our easy assumptions about what is natural, what is right, and what progress among humans might actually exist. Frankly, I was struck by the level of harmony and refined social coordination that pervaded the !Kung society before the arrival of the Europeans severely disrupted their world.

When I watched that hunting party track a giraffe for several days before finally killing it with their spears in the ethnographic documentary film I mentioned earlier, the hunters' coordination and control involved was clear. The relations among the four hunters were fascinating to me, especially in light of a later analysis by historian William Irwin Thompson (1971). Each member had a specific role to play in coordination and decision-making and action during the hunt, although the roles were quite informal and not bound by any discernible hierarchy. Their complementary opposite roles were, chief, warrior, medicine man, and clown.

That might sound odd, but these are just labels we have applied to capture the essence of these individuals' contributions to the cohesion of the group that made for a successful hunt and allowed the group to function so well under extremely difficult conditions. Simply put, the chief was the ultimate decision maker who listened to the others, the warrior was the technical expert with weapons and hunting techniques, the medicine man provided sage philosophical advice and cultural context for this critical mission, and the clown offered comic relief. It all added up to one harmonious operation aided by the complementary opposite roles played by the men: chief vs. warrior and medicine man vs. clown. Thompson likened these roles as functional precursors to institutions with similar operations and complementarities.

The narrow perspective of our modern consumer culture would view the !Kung hunter-gatherers of the Kalahari Desert in Southern Africa as destitute "savages," barely surviving from day to day. After all, how could always living on the edge of survival be the good life? On the other hand, living within the modern consumer bubble gives the false impression of being protected from the risks of living in the natural world. Yet, we do live near the edge of survival, actually, the edge is much closer than you thought, and it is getting closer by the day.

As I have sometimes put it, on any day I venture out in public, a truck could hit me and I would be gone in an instant. Countless such risks exist, many of them collective and unrecognized. For example, many biochemical risks are realized in increased rates of what some call modern diseases. We may never be directly aware of many modern hazards until their consequences hit us—even then we may not know the source. Safety has become a partly abstract topic in the industrial-consumer culture while the entirety of industrial civilization moves ever closer to the edge of chaos and collapse.

For me, the !Kung represented something far more interesting than just another ethnographic study of a remote tribe. The !Kung provide a clear example of successfully living in harmony with an unforgiving ecosystem. It is all about the context. Widespread "affluenza" (the disease of addiction to affluent comfort) prevents many Americans from seeing how precariously close to the edge we all really are. Our culture lives at the edge of illusion, one foot in reality, the other in consumer fantasies or, far too often, conspiracy theories to explain whatever our Facebook friends fear the most. The consumer bubble has a foggy sheen. The COVID-19 pandemic lifted that fog a bit for some, while others remained in political denial that anything in Nature or society could possibly burst that bubble. Therefore, "It must be a conspiracy! Arrest the usual suspects!"

In what seems like another world, flying in tight formation in an F-16 Fighting Falcon jet-fighter aircraft is living on the edge of survival, though in a very different way. Coordinated high-speed formation flying condenses survival risk down to the microsecond. Rather than performance on a long tedious yet dangerous trek, split-second precision is the measure of success. Failure is not an option because death is the immediate consequence.

Timing and precision are everything. The documentary I watched about the Air Force Thunderbird aerobatic team demonstrated the personal dedication, rigid training, and extreme coordination among its members. These members execute their mission to entertain the public by displaying their highly refined skills in aerobatic formation flying.

Air Force Thunderbirds and the Navy Blue Angels coordinated achievement of executing such precise maneuvers provides the personal control necessary to respond to the complexities of actual combat. That, of course, is the whole point of military discipline for those who understand it as Colonel Boyd did. Precise personal discipline in coordination with others, which maximizes social control, is what being a top gun really means—the exact opposite of John Wayne's shoot-from-the-hip unpredictable "rugged American individualism." And, oh yes, forget about Tom Cruise's ego.

The real world is always much messier than depicted in action hero movies. Unfortunately, police officers have to endure the burden of long stretches of boredom, never knowing when they will suddenly find themselves in a genuine combat situation—an ambush or a fire-fight at a traffic stop. Under such ambiguity, it becomes difficult not to overreact. To protect and serve is far more complex than it sounds.

Unlike the linear thinking behind traditional military and police theories, John Boyd's model recognized the nonlinear processes involved in the dynamics of combat (Coram 2002). Yet, the dynamics of creation and destruction apply in all complex adaptive systems and their environments, from hyperactive episodes such as aerial combat or street confrontations to a business meeting or everyday police-citizen contacts, even to the functioning of entire societies. Police work is not military combat, but a creative effort may suddenly transform into an event driven by destruction.

As the growing techno-industrial economy recreates itself by extracting materials and producing waste in ever greater volumes, on its ill-fated path into the Anthropocene, it rapidly generates new conditions of all kinds. This creates the growing need for rapid personal and societal changes involving responses of both destruction and creation that we can barely imagine but must prepare to invent and carry out very soon. The latest new twist at this writing is artificial intelligence (AI), especially large language models that mimic human writing by "learning" from vast data sets. The applications of

AI are many, and many are controversial for good reason, but that is another story.

One of the biggest concerns about AI—aside from its ethical and cultural issues—is its huge consumption of electricity to process its giant data sets. Some forecast severe energy shortages due to widespread AI deployment, just when we need to reduce energy consumption. As we approach the end of the Industrial Era as we know it, we move into uncharted territory.

The industrial-consumer culture is no more prepared for this than a !Kung tribesman prepared to fly air combat missions or an F-16 pilot prepared to survive over time, with a spear and a dibble stick in the Kalahari Desert. We live by industrial-consumer habits, but the near future portends a very different cultural as well as ecological landscape.

Boyd did not follow a script. He responded creatively to rapidly changing dynamics by applying his situational hyper-awareness, highly refined skills, and ability to see the big picture as it unfolded and changed. It may seem strange, but the same kind of dynamic dilemma presents itself to both the current hierarchy of the globalized political economy and the formalisms of everyday urban law enforce- ment, government legislation, and policy formation.

Some groups seem entirely fluid and open to innovation and new members. Others are rigid, closed to outsiders, and intolerant of deviance of any kind. Consider the differences between the Taliban and a jazz quartet. Coordination is key in both, but creativity is the essence of only one. In a rapidly changing environment, rigid ideology will fail.

Some police departments become dysfunctional because their hierarchical worldview causes their officers to be unable to respond creatively to the dynamics of the complex systems they must engage— be they communities, gangs, deranged addicts, the mentally disturbed, or the dynamics of escalation in any conflict situation. That is why I advocate vastly more advanced vetting and training for police recruits as well as high pay with high-performance expectations, along with a more dynamic model of public engagement.

In the US especially, we value our personal freedom. At the same time, we expect the security that a stable social order provides. We often expect others to conform to norms we imagine are universal, but we do not always hold ourselves to quite the same standards. Today, we face the necessity of exercising creative destruction to transform society in unprecedented ways to achieve a rapid reduction of carbon emissions and the end of ecosystem destruction to create a livable place for us humans in the early Anthropocene.

The role of law enforcement that our culture so far envisions will be inadequate or even problematic under the changing social conditions of the emerging global transformation, both ecological and societal. The role of change agents in the context of holding it together into the Anthropocene, both within the institutions of social control and among social movements for Earth-System stability, may be best modeled after John Boyd's system-dynamics approach to creation and destruction. Serious social change simply does not happen in the context of business as usual. In that context, law enforcement becomes either an enhancement to progress or a impairment to social control.

Social Control and Dilemmas of Hierarchy and Force

In small indigenous communities and tribes, there is no such thing as law enforcement, just norms grounded in cultural consensus. Social control emerges from the essential elements of culture, not from the top of a social hierarchy. Those elements are internal norms and beliefs about what is right or not, what I need or not, and what we need to control behavior in everyone's interest. Here is what another anthropologist observed among the !Kung:

"Since hunter-gatherers are often hungry, one might imagine that food theft would be a daily problem. Like other people living in small-scale egalitarian societies, they have no police or any other kind of authority … Men might return to camp at any time, alone or in a small group. Many of the foods a woman cooks are edible raw, so they could be eaten before, during, or after the cooking process. If a man returns from the bush feeling hungry and has no one to cook for him, he might be tempted to ask a woman for some food—or even simply take it—rather than doing his own cooking.

Yet such tactics are rare. The relaxed atmosphere Lorna Marshall described for the !Kung is due to a system that keeps the peace at mealtimes among hunter-gatherers and other small-scale societies. The system consists of strong cultural norms." (Wrangham 2009:160-161)

In small societies where the "rational-legal" form of authority has not taken hold and hierarchy is minimal and consensus is high, the force of mutually agreed norms is enough to minimize bad behavior. From childhood on, people have internalized the norms of the group. Most members accept them and behave accordingly as a matter of honor.

If someone violates a norm, then shaming or public ridicule will do the trick. This can work within small groups in modern complex societies, too, but not so much on the larger scale of industrial society as a whole. Aside from their origins in the enforcement of monarchic control over subordinate populations, the role of law enforcement institutions and the individual police officer is to enforce norms written into law. Nevertheless, non-legal norms play a role too.

Modern societies expect too much from and offer too little support for their law enforcement institutions and officers. Admonitions about law and order are not enough. As with teachers and other public service professionals, a pay scale that reflects their importance to society, along with very high standards of admission and performance, are necessary yet rarely achieved. Although these necessities are not forthcoming, at the same time, people are shocked when things go wrong.

Complex industrial societies, in part because of the elaboration of complex rational-legal institutions, cannot maintain a uniform normative order within their entire populations via interpersonal relations of trust and commitment. If anything, they are just too complex. Too many relations operate at too great a social distance. In addition, the diversity of human groups and their different standing within such societies make the kind of normative order found among mostly egalitarian indigenous peoples nearly impossible to achieve or maintain at the societal or even community level.

Rational-legal normative structures focus on the requirements of the larger economic system. Many laws or regulations have little to do with the everyday lives of the people except in relation to the rules of the organizations with which they engage. Such normative systems connect less directly with people's everyday lives than with their institutional roles. At the same time, a whole range of laws attempt to control public behavior, including everything from traffic regulations to commercial robbery, shoplifting, assault, murder, and other felonies, many of which are as likely to be economic as personal. Law enforcement, more than cultural norms, constrains food theft to the extent that it can.

At the same time, extreme conditions of personal and community deprivation cause high levels of stress and, therefore, high levels of "self-medication," interpersonal conflict, domestic abuse, and impulsive violence. Domination and deprivation breed dysfunction. The societal/political response is to blame the victim and look no further, guaranteeing the perpetuation of the problem of social disorder. Political authorities hand law enforcement institutions an impossible task: solving a myriad of social problems and offering inadequate resources, training, and compensation to optimize the role of law enforcement in society.

Industrial civilization consists of hierarchies of hierarchies of mostly complex economic and political institutions. Social control outside of families and neighborhoods operates mostly based on formal laws and regulations. Even communities formalize their structure as rational-legal jurisdictions. The pervasiveness of such formal control mechanisms everywhere in society makes the maintenance of family and community bonds and the norms that are part of family and community life secondary to the requirements of the larger institutional structure. Therefore, it becomes more difficult to maintain.

Family and community are at the bottom of the national and regional complex political-economic hierarchy, nested within and subordinated to higher levels of authority. Some political and economic higher authority subordinates every family and every community to its control, even as they operate at a certain social distance. In that context, they contrast severely with the integral bonds and norms that bind individuals to their families and tribes within hunter-gatherer and similar indigenous, often egalitarian, cultures. Yet, everyone takes for granted many norms. Except for the most egregious drunk driver, everyone drives on the right side of the road.

Because of these complexities of modern life, social control exists within each political jurisdiction through the deployment of economic incentives and punishments through regulations, tax credits, and penalties like fines or legal enforcement. Police enforce "the law" in localities, focusing mostly on publicly observable behavior. Here, the threat of force usually involves arrest and prosecution, leading to incarceration.

Police also respond to various domestic "disturbances," larcenies, reports of physical violence, and other alleged violations of law, and in each case, face some degree of danger and situational ambiguity. They have a great deal of discretion to interpret violations and to apply various levels of force, depending on how a situation evolves, sometimes very rapidly, sometimes including lethal force. Ultimately, the mandate of law enforcement is to maintain the formal political-economic hierarchy of society, as they perceive it, by force if necessary, and to subdue anyone who attempts to violate their authority.

Law enforcement often emphasizes social control less within a framework of social relations than it is about reference to institutionalized control of populations to ensure they conform to the strictures of the "rational-legal" order as defined by the hierarchy of authority. This is one major factor that makes policing so difficult and often ambiguous. The other factor, of course, is the unknown level of threat to officers' own safety, the ambiguity of which can result in overreaction and excessive use of force. Here, vetting and training are key and are often neglected, in part due to insufficient public funding.

But again, our modern condition consists of a hierarchy of elites and various forms of subordination of social, class, racial, and ethnic groups, each of which endures forms of exploitation, degradation, and exclusion. Law enforcement institutions are deeply embedded in that hierarchy. The little chance we have to build an ecological civilization will be lost if we fail to overcome the deeply engrained hierarchy of domination and subordination, both political and economic, as well as cultural. That is why the potential of transcommunality (Childs 2023) for developing the kind of resistance to the terminal trajectory of the global political economy of growth is so important.

Unity through diversity can thrive only by instilling a culture of respect. If we can achieve that, then a high level of social mobilization will be able to respond effectively to the global threat of destabilization of ecosystems, climate, and society itself. A growing social movement grounded in transcommunality can exert public pressure on political and financial elites to force decisions in the interests of human and planetary wellbeing rather than endless capital accumulation by the few.

The role of law enforcement that our culture so far envisions will be ineffective or even problematic under the changing social conditions of the emerging global transformation, both ecological and societal. The role of change agents in the context of "holding it together" into the Anthropocene, both within the institutions of social control and among social movements for Earth System stability, may be usefully modeled after John Boyd's system dynamics approach to creation and destruction. Serious social change simply does not happen in the context of "business as usual."

The societal transformation we need will not happen without a massive social movement to exert the necessary public pressure on politicians and plutocrats alike. A social movement of the scale and complexity needed is unlikely to be formed without building an ethic of transcommunality among the diverse groups and members needed to participate in a global social movement for climate/ecological/societal transformation.

Economics, Policing, and Community Needs

Today, video technology documents many of the relationships between economic deprivation, law, and policing, but the resulting knowledge rarely achieves clear-cut influence on public policy. The culture treats crime in its simplest cultural frame as if a violation of law is just that, nothing more. However, crime is complex at any level, whether it occurs in the highest reaches of political or economic power or among the poorest citizens in society. Body cams and dash-cams record police-citizen interactions, sometimes without much context.

The context may be beyond the camera's reach. Nevertheless, "good cops" usually welcome documentation because they recognize that it can validate proper police procedure.

The implications of law enforcement as we know it, or perhaps as we do not know, for the future of a society that is already beginning to undergo a New Great Transformation are profound and unfamiliar. As we enter the early stages of the Anthropocene Epoch and all the unknown elements it portends, we must sort out our understanding of how economic change, law enforcement, and the changing needs of the public interact with one another. How does law enforcement fit into a rapidly changing societal environment? That may be more difficult to answer than it seems.

Law enforcement, you might say, consists of formalized efforts to assert or maintain social control. In ordinary everyday life, society achieves social control through consensus, consent, and general compliance with norms of behavior. People stay within the bounds of established norms in response to different levels of acceptance. If they are a party to the social consensus (that is, they have internalized the norms), acceptance is virtually automatic because the norms are integral to their worldview and personal identity.

However, if a norm or legal requirement does not entirely fit within their personal belief system, yet their general belief in the acceptance of norms is stronger, most folks will consent to that norm and behave accordingly. Now, mere compliance is a somewhat different matter. If a person does not accept a norm, then they may or may not comply with it for other reasons or for no reason at all. Compliance for them may be a matter of calculating the cost of violating the norm and/or any reward that may result from compliance. However, impulse also plays a role.

It may simply be a matter of weak impulse control or a broader personal indifference to social norms in general. A person with little impulse control will often not calculate the costs of violating a norm or consider the benefits of compliance until after the fact. In the case of sociopaths of various stripes—murderers, rapists, and political grifters—behavior control rests solely on self-indulgence and the degree to which they may calculate the threat of exposure to their immoral or criminal behavior. They lack an internal normative structure (moral center). Some characterize their decisions or political policies as strictly transactional.

For the most part, people internalize the norms of their family and the immediate community in which they grow up, although not all do so. In hunter-gatherer societies like the !Kung, consensus tends to be high because the survival of the group requires a high degree of consensus, cooperation, and mutual aid. Often, alternatives simply do not exist. A high degree of consistency is present in such contexts. Everyone knows their role and practices it in coordination with others. No "alternative lifestyles" present themselves within the confines of their habitat or culture. Such notions are distinctly limited to industrial-consumer societies.

In modern industrial societies, a number of competing subcultures typically exist side by side. We call that pluralism. A dominant culture sets the rules, and the various sub-groups try to maintain their own cultural values while complying with the dominant society's norms or, at least, the laws. In such non-indigenous societies, cultural conflict is bound to occur.

Typically, many members of the dominant group resent the very existence of different cultures in their midst, especially groups defined by race, even if their members fully comply with the dominant norms and laws. The contemporary example of this in the US is the so-called "culture wars," although a number of other factors are involved. Political demagogues exploit the fear and resentments felt by members of the dominant culture who experience status anxiety and/or economic loss due, for example, in the case of older White men in the US, to the loss of well-paid blue-collar manufacturing jobs to overseas outsourcing or automation. Society's most vulnerable groups become scapegoats.

Industrial-consumer economies tie individuals strongly to their separate struggle to find a viable source of income and subsistence, which tends to isolate them socially and psychologically from family and community. Traditional social bonds weaken, and associated normative structures become tenuous. Alienation and dissociation follow, often resulting in social chaos of one kind or another. Insecurity and status anxiety may result in various forms of acting out. Practices of discrimination based on race, class, and culture produce an ambiguous situation in terms of both norms and law. Compliance with both norms and laws becomes more situational, more transactional, and more ambiguous. Many of the more successful members of the dominant group may not even be aware, from their suburban vantage point, of the fact that the subordinate group is a target of discrimination, even hate. They traffic in social illusion, enabling themselves to remain comfortable with their (mostly White) privilege.

Holding it together under these conditions becomes tenuous at best. Social cohesion results primarily from interpersonal relations that benefit all involved. This is true from the bottom to the top of the societal hierarchy. The January 2023 struggle to elect a speaker of the US House of Representatives reflected both interpersonal struggles among the Republican Party Congressional Caucus and the breakdown of societal norms. This breakdown occurred both outside and within the collection of individuals elected ostensibly to represent sectors of the American people. Clearly, something else was going on.

The new extremist element in the GOP reflected a parallel expression of chaos in the general social order. This kind of social chaos puts the normative structure of society itself at risk in several ways. Narcissists and sociopaths in political positions tend to exploit such chaos by making extreme demands on the system to gain as much power and attention as possible, as if they were the ultimate authority. That game sometimes works, especially when they pander to the resentments and demands of the very power elites who fund their extremist views and actions. When numerous members of Congress conspire with insurrectionists to prevent the traditional peaceful handing over of power to the newly elected president, we see a deeper fragility of democratic institutions.

In the context of a capital-intensive economy where jobs, income, and security appear increasingly to have less and less stability, the economy fails to provide for community needs, even as the phantom wealth of financial and corporate elites grows to obscene levels. Growing poverty and economic exclusion are strong predictors of societal chaos, street crime, and related police violence. This results, in part, from the societal failure to grasp the most fundamental facts of crime and punishment.

A society holding it together does not result from more aggressive crime control, although that can become necessary in some form. Instead, genuine social order results from social inclusion and mutual support in many forms throughout society. These processes are the most effective sources of social control. They can minimize social dysfunction and crime yet seem further and further from our reach. However, autocratic attempts to force a new hierarchical power on democratic institutions threaten civil order. If the center does not hold, then some kind of societal transformation will happen, one way or the other, for better or worse, in the coming decades. The direction of change remains an open question.

The one burning question that remains is if we can hold it together enough to take charge of the changes around us and create a new social order that sustains our habitat and ourselves. Today's deteriorating social conditions resulting from environmental degradation require us to engage in massive ecosystem restoration. We also need to reorganize our economies around the values of human wellbeing and mutual support instead of the dead end of capital accumulation by elites and the degradation of life conditions for everyone else. To hold it together in the near future, we will need a resurgence of social bonding and mutual aid in the struggle to form a new viable ecological civilization.

Community Control, Law, and the Converging Crises

One of the most fundamental changes in the global transition from indigenous societies to agrarian and then to industrial-consumer economies was the shift of social control from interpersonal relations within communities (local indigenous self-governing groups) to increasingly complex economic institutions. This shift ultimately culminated in the formation of nations. Finally, a transnational matrix of corporations and the governments that sustain them politically became the dominant determinant of social control.

That is a mouthful, I realize, but this ultimately dehumanizing process some praise as "globalization" has engulfed the whole world. Colonialism, imperialism, then economic and financial domination were the instruments by which political-economic dominators took social control away from communities. Both internally and in the Global South, the global complex of extractive international institutions of the Global North came to dominate the world. That has had countless ramifications for the lives of people everywhere.

Hierarchy has won its struggle with community, at least for now. The dominance of the world by complex institutions has taken centralization of power about as far as it can go. I say the center cannot hold because multiple signs of instability, decay, and decline are in evidence, from international financial crises to increasingly extreme inequities in the distribution of income and wealth that cause severe poverty and community dysfunction. Societal decline is happening in the context of an accelerating convergence of the multiple crises of ecological, climate, and other threats to the survival of diverse societies around the planet.

The steady growth of industrial-consumer economies and their international financial integration, so-called "globalization," has reached its peak and has now begun to falter. Governments find it more difficult to govern, and they increasingly fail to provide for the wellbeing of their people. They are ever more unlikely to hold it together. However, what is far more important for the people and the planet is if we can transform, or in most cases replace, these archaic bastions of "modernity."

We must replace the faltering ecosystem-destroying and climate-destabilizing dehumanized hierarchy of globalized institutions with viable ecological communities that can sustain themselves within the limits of the local-regional ecosystems that they must also restore and regenerate. In that context, we must ask what the role of law and order in an increasingly chaotic world is—a very difficult question at every level, from local to international.

As industrial-consumer economies become more inequitable, with income and wealth concentrating more and more at the top of the hierarchy, which Marjorie Kelly (2023) calls "wealth supremacy," the role of law enforcement agencies becomes more controversial and their effectiveness more difficult to achieve. At the same time, the emergence of new levels of both societal chaos and inconsistent police practices has elicited very little political motivation to adopt new strategies on the part of law enforcement agencies as societal conditions become more unstable.

Mixed messages abound. Political decisions to not pursue perpetrators of minor crimes, such as shoplifting, were followed by a surge of smash-and-grab group robberies. Several perpetrators simply grab several hundred dollars' worth of merchandise and walk out of the store. Stores discourage employees from intervening, out of management fear of liability for injury. Violence increasingly accompanies such crimes.

On observing a gang of shoplifters loading up booty in a big box store, one off-duty sheriff confronted them demanding that they replace the merchandise. All but one dropped the goods and ran. The remaining shoplifter became belligerent, and the sheriff took him down. What does this tell us about social control? Plenty. The sheriff responded to the situation in terms of his moral code and personal identity. The thieves had no such code. In general, social control has become more tenuous than ever, with law enforcement put in an increasingly difficult position. As the sheriff in this case put it, "I couldn't just stand there and do nothing!" News clips rightly lauded this officer's individual bravery and commitment to moral principles. But where is the public-policy meat?

As society transforms in the coming decades, law enforcement already finds itself in an increasingly ambiguous position. Larger institutional forces have pressured police to become indistinguishable from an occupying military force, particularly in downtrodden urban neighborhoods with high crime rates. Without integrating the most vulnerable populations into a viable economy, the authority of the managed pseudo-democracy of the corporate state will only degrade into further routine destruction, leaving law enforcement in an even more untenable position. Unless radically transformed, the degraded civic culture will leave law enforcement culture and institutions increasingly and dangerously unprepared to deal with the growing chaos.

The so-called "deep state," with its instrumental institutions of empire, will remain mostly intact (Lofgren 2016) as long as civilization's collapse remains avoidable. (Here, I am not talking about the conspiracy theorists' Q-anon-inspired fantasy of a government plot to overthrow our freedoms, etc., but the complex institutional structure that parallels the political economy.) Politician-encouraged militarized local law enforcement practices will likely further entrench themselves as long as political institutions have little or no grasp of the causes of growing chaos.

Increasing the distance of law enforcement from the public it is sworn to protect and serve shifts their relations from mutual support to dominator-subordinate relations of mutual resentment. In stark contrast, if we are to sustain civil society, new or reformed institutions seeking ecological and social integration must replace those of a failing empire of destructive globalization. Under such a mandate, law enforcement might take on a whole new role. But how can we accomplish either of these transformations?

Taking diverse restorative actions where we live can build a new form of social control by cooperative action that sustains human wellbeing, overcoming fear and destruction with social solidarity. That could offer police a new, positive mandate. Becoming grounded in community authority is necessary for viable law enforcement and ecological restoration.

How to achieve such a societal transformation toward organic social control remains an extremely difficult conundrum. Really holding it together will depend on the establishment of mutual trust. In organic, integrated communities, much social control takes care of itself because of all of the elements, both humans and flora/fauna. They maintain stable relations of mutual support with one another as balanced local-regional, social-ecological systems. Law enforcement that has community authority can strongly assert consensual social control.

The trick is to get there from here. That would require both community and police cultures to change—no easy task. But to achieve community-police social solidarity through mutual support would lead to much stronger social control as the context for a new kind of freedom grounded in mutual respect and cooperation. That is very unlikely to happen until the growing extreme economic disparities between political-financial elites and everyone else are severely reduced.

If communities can achieve ecological viability, law enforcement could move closer to an affirmation of a social order grounded in community sovereignty and social system integrity. Most police officers would much prefer to participate in viable community sovereignty than have to deal with the chaos they must face now as an outside force. In that context, law enforcement could integrate its operations with community needs under the increasingly difficult environmental conditions of the Anthropocene, no longer floundering in a lost cause.

CHAPTER EIGHT

Empire and Terror: Out of Control

EMPIRES, THE TERROR they commit, and the terrors their victims return to them have been around for millennia. However, modern technology and organization have taken both to entirely new levels of power and destruction. The 9/11/2021 destruction of the Twin Towers in New York City and associated damage to the Pentagon by a rag-tag collection of Saudi nationals—ironically using America's high-tech airliners as weapons—became an icon of terrorist response to the global corporate empire of the industrial-consumer economy of growth led by the US.

The Twin Towers were the symbol of American financial dominance of the world. Subsequent US' wars of choice in Iraq, Afghanistan, and elsewhere reflected the same struggle of empire and insurgency that characterized the relations of colonial conquest, then the imperial dominance by the Global North and the resistance by the peoples of the Global South. Rebels, insurgents, and terrorists have always had an advantage. They could attack the dominant power anywhere, any time, but now they have modern weapons of war. That is why the American colonial rebels won the Revolutionary War.

The hierarchy of relations of domination and subordination among nations is not structurally different from the relations of individuals in hierarchical groups. Democratic relations are far more conducive to peace and less prone to the extremes of violence. Democracy is grounded in the value and equitable power shared among all its members. Empires generally produce riches for the leaders of the empire at first by exerting control over their victims. Eventually, though, they tend to generate resistance, rebellion, and finally insurgent terror, usually around the time they reach their peak and begin to decline.

We could say the same about the British failure to hold onto its colonies in North America and elsewhere. Once economic exploitation and political oppression reached a breaking point, rebellion and revolution ensued. Today, the main difference is that the technologies and tactics of empires and their increasingly rebellious subordinate states and non-state actors are far more complex than ever before, as are their consequences for people and the planet.

As the end of the Industrial Era approaches, the continuing struggles between empire and terror make resolving the destabilization of the Earth System that much more difficult. How can we hold it together in completely new ways when the old conflicts of power lock us into the mentality and behaviors of the past? What we need most is the exact opposite: high levels of international cooperation and compassion in transforming the terminal world economy into a viable ecological civilization.

Empire of Growth, Overshoot, and Ruin

From its earliest beginnings in world exploration, conquest, and colonization, the need to expand national power beyond one's borders drove the modern global political economy. Born in colonial revolt and insurgent war against tyrannical British rule, the US expanded across the continent by means of territorial invasion, confiscation, occupation, and the subjugation and extermination of native populations. Along the way, it conquered the parts of Mexico that would become Arizona, New Mexico, California, and Colorado.

During the Westward expansion, the US industrial system was able to supply almost all its expanding needs for raw materials and energy from within its borders and conquered territories, including minerals and coal, and later oil and gas. By the beginning of the twentieth century, however, the industrial nations of Europe and North America realized that control of the world's oil fields would become necessary (due to the ascendance of fossil fuels as the primary energy source for industrial operations and growth). By economic power and military threat, the industrializing nations could obtain the raw materials and oil needed for continued expansion.

Gradually, the colonizing nations relinquished formal political-military administration as the means of control by instead dominating world trade and finance—a more efficient gambit. European nations began exerting power over former colonies economically instead of militarily. Growing industrial capacity became the engine of the empire. The US picked up some territories and "protectorates," mostly in the Caribbean and the Pacific, and dominated Latin American nations both politically and economically but not by traditional colonial administration.

The US was able to gain access to various resources through economic influence and political pressure, occasionally supplemented by Marine Corps invasions and occupations such as in Cuba, Guatemala, Haiti, and the Philippines. Eventually, economic hit men who were (are?) hybrid agents of both corporate financial and state power enforced US economic domination, mostly in Latin America and the Pacific, occasionally augmented by secret political assassinations (Perkins 2004).

The idea of manifest destiny framed both Westward expansion and, later, the self-righteous domination of all the Americas from the mid-nineteenth century on—it remains implicit in US foreign policy today. Retired US Marine Corps Major General Smedley D. Butler, a two-time Medal of Honor recipient, wrote a scathing small book, *War is a Racket* (1935). Butler described how the only benefit of the wars he and his men prosecuted was the profits US business interests gained from those wars. Butler concluded that the heroism and sacrifice of the military and the human suffering of war benefited corporate profits but not humanity. And the beat goes on.

By its engagement in World Wars I and II, the US built its military strength and political influence in the world and its industrial might well beyond that of any other nation. It became the first fully industrialized global empire in history, backed by the world's most powerful military. The emerging Soviet Empire competed on economic, ideological, and military grounds but could not wield its centralized bureaucratic power effectively to keep up with the more freewheeling industrial might of US state-supported corporate capitalism. With the eventual collapse of the Soviet Union, the US became, by default, the one remaining "superpower." Or so thought most analysts of international relations. Gradually, however, things became much more complicated.

Building and sustaining a modern industrial empire, and the continual economic growth and the debt-based capital accumulation it demands, means that it needs access to more and more resources. To extract both materials and energy and to create new markets for its growth requires effective control over territories and governments, as well as a steady supply of capital. Classical economic theory framed the growth of the British Empire glowingly and, later, the even more powerful US economy of growth but conveniently ignored the human costs of empire.

Major US governmental subsidies, such as funding Westward expansion of the railroads, were key drivers of early economic growth. State funding was a prime engine of that growth, in contradiction to the cultural myth that rugged individual entrepreneurialism did it all. The combination of private investment and government subsidies stimulated much of the exploitation of resources and industrial development that drove nineteenth-century economic growth into the twentieth century.

Neoliberal corporate economics dominated the US political economy throughout the twentieth century and the first quarter of the twenty-first century. Then, the system of endless growth began to run up against environmental, economic, and societal obstacles. Many empires have ascended and fallen, yet their leaders never expected an end to their power. That has not changed.

From Alexander the Great of Macedon and Julius Caesar crossing the Rubicon from Adolph Hitler's delusions of world domination by an imaginary master race, the story continued. Donald J. Trump's delusional tower of achieving and retaining anti-democratic autocratic power followed Dick Cheney's fantasies of controlling Middle East oil. Each lived in the illusion of his personal power over a perpetual empire. However, all bad things must end, even if followed by another bad thing.

Any system that continues to grow requires constant energy inputs to sustain that growth. In Nature, growth occurs in relation to the life cycle—from birth through accelerated growth followed by slowing growth. Then growth stops; gradual but accelerating breakdown and death eventually follow in the natural course of life. Diverse forms of natural growth consistently follow an S or logistic sigmoid pattern called the logistic curve (Figure 3 below). It is fascinating to know that the growth of a tree follows the same basic pattern as many other natural cycles of life.

Scientists have applied variations of the logistic curve to describe many processes. Chemical reactions, various tumors, the diffusion of innovations in an economy, and the population growth of any species in a particular ecological niche all display this basic pattern. In every case, growth gradually slows, then stops. If ecological stasis occurs, the population remains steady as individuals come and go, as shown in Figure 3 below. Suppose a population or an empire overshoots the capacity of its habitat (including its client states) to sustain it. If that happens, rapid collapse ensues, as illustrated by the graph of a Seneca collapse shown in Figure 2, Chapter 4.

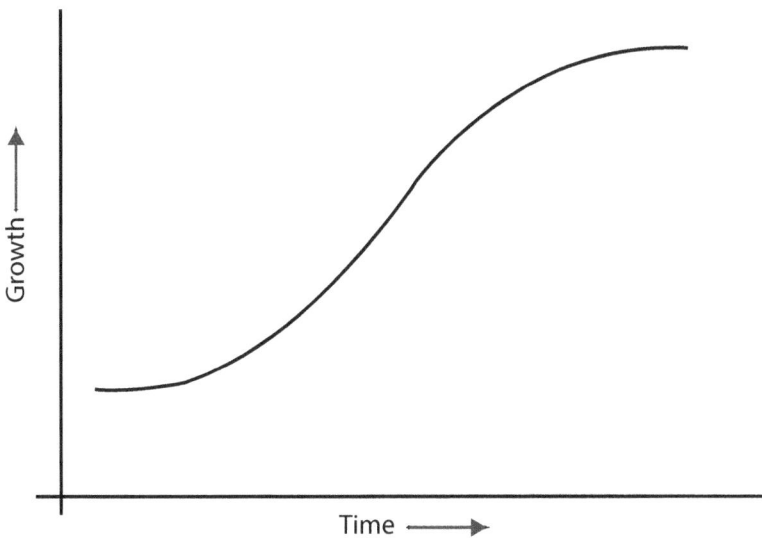

Figure 3. The Logistic Sigmoid Function

Rendered by Nick Zelinger. For more information, see Wikipedia: https://en.wikipedia.org/wiki/Logistic_function

The wide applicability of the logistic curve shown in Figure 3 reflects the natural growth cycle of systems in finite environments (before old age and death), not its individual members. Growth, maturity, and decline tend to reflect the relationship of any system to the resources and energy available in its environment and its own internal life cycle, including death. We can even describe the life cycle of empires by the logistic curve, with one important exception.

Many organisms live on for quite a while after growing to maturity, then very gradually decline and die—such as the Sequoia trees of Northern California, which live for hundreds, even thousands, of years. Its life cycle will fit the logistic function that would continue flat until it slowly dies off. If a species is part of a stable ecological system, it can reach a maximum population size and continue to exist among others, as long as it remains in balance with the other species in that ecosystem.

After reaching its maximum height, a Sequoia may live for hundreds of years or more unless the forest ecosystem is severely disturbed. If a population of elk overshoots the capacity of its habitat (its ecosystem) to support growing numbers in the absence of wolves, then it is likely to experience a Seneca-like collapse. Such changes can affect many other species in the ecosystem too.

Once the growth of an empire has achieved its peak, slowed, or stopped growing, it tends to stagnate, destabilize, and begin to decline. Usually for a combination of reasons, ranging from internal corruption to inability to sustain costs and many others, it begins to decline, then the decline accelerates, resulting in a reversal of the logistic curve but with a steeper decline. The decline phase is usually much more rapid than the growth phase. What might have been a symmetrical bell-shaped curve takes the skewed form of a Seneca Collapse, described by Figure 2 in Chapter 1. This pattern of decline and collapse parallels the rapid decline and death of an organism due to cancer or other systemic failure.

Depending on the species, once a tree reaches maturity, it may live in stability for a long time before slowly dying. The great California Sequoia can live for many centuries, finally dying off very slowly from the periphery to the center. In contrast, the often-rapid decline of empires following maturity may even be catastrophic. Numerous historical cases attest to the social illusions that lead to societal collapse because leaders and elites ignored the reality of the environmental requirements for societal stability and survival (Diamond 2005; Tainter 1988). Other complications, such as threats from rival empires, resistance from subject peoples, or difficulties in controlling an extended hierarchy may accelerate their collapse.

The Roman Empire may be the most obvious and widely known example of such decline and fall, most famously documented by Edward Gibbon ([1776] 1995). Growth naturally slows and then stops as an empire approaches the carrying capacity of significant sectors of its extensive environment. When an empire grows too big, it becomes unwieldy and more difficult to manage or sustain. Diverse factors, such as military outsourcing to mercenaries (contractors in current euphemistic parlance), internal struggles for power, or financial crises, contribute to its growing weakness.

Internecine struggles typically ensue within maturing empires, along with growing corruption (Chayes 2015) and a destabilizing political system. Sound familiar? The Roman Empire, at its peak, controlled a territory that now contains nearly fifty modern nations. Scholars have debated the causes of its decline and fall for centuries. However, like most empires, its decline was relatively rapid. We can expect the same pattern of decline and collapse of today's globalized empire of economic growth, but, consistent with history, its "leaders" do not.

We can draw parallels between classical empires and the unprecedented international economic and military global reach of the US-led political-economic empire today. Yet, important differences exist. Not long after the 9/11 terrorist attacks on the World Trade Center in New York City and the Pentagon in Washington, DC, Chalmers Johnson reported on the US military global reach. The US military by then had established over 725 known foreign bases in 153 countries, with hundreds of thousands of military personnel deployed around the world (Johnson 2004, 154). Johnson's estimate did not include the unknown number of secret bases or associated clandestine operations and "black sites." Most other nations have no foreign military bases at all, and the US established most of its bases long before the 9/11 attacks. The modern empire of economic globalization exerts the power of economic, political, military, and clandestine operations worldwide. The unanticipated consequences of such operations are damaging to the empire itself. Chalmers Johnson explained some of these in an earlier book, *Blowback* (2000). Johnson was a highly respected historian and Japan expert and had been a consultant for the US intelligence services. He built his critical assessments of the American empire on deep personal experience within his subject matter.

Modern colonialism began when the earliest explorations by Western conquistadors employed new military and navigational technologies—gunpowder and the sextant, in particular. Their ruthlessness in satisfying their greed often reached the ferocity of modern jihadists seeking millennialist redemption. Vasco da Gama and his successors used indiscriminate and wanton violence to establish the first global empire. The Portuguese set up trading outposts on the west and east coasts of Africa and established trade routes via the Indian Ocean, attempting to usurp the Venetian dominance of European trade with Asia (Crowley 2015).

Ever since, the most powerful nations have subjugated and exterminated diverse peoples around the world in the name of their vision of progress—military, political, and economic domination. The mandate of the earliest European explorers was to find riches and return them to the monarchs who funded their expeditions. The invaders routinely enslaved or exterminated native populations to build colonies and further exploit their lands. The modern empire of corporate world domination differs mainly in its advanced levels of technology and organization and in its politically correct rationalizations for its oppression of others.

Because of international financial integration and military cooperation, the modern industrialized global empire is, in many respects, a multinational empire. The US leads the global-industrial empire today with variable participation by its allies in the Global North–primarily the European industrial nations of NATO. As with the ill-advised and ill-fated Vietnam undertaking, much of the motivation for the adventurism of empire today seems as much ideological and political as economic. The "defense" industry, however, relies heavily on cost-plus contracts that guarantee a handsome profit. Military contractors thus have incentives to lobby for invasions and occupations and to inflate costs, which contribute much to the overshoot of the empire.

Military efforts to dominate targeted nations have gained much less than hoped in terms of the spoils of empire, such as control over industrial raw materials and oil. Cheney and Bush never realized their dreams of plundering Iraq's oil wealth. By destroying the civil and governmental infrastructure, they caused their own failure to gain the social control needed to extract it. Yet, vast profits have accrued to the military-industrial complex, which sells evermore equipment and arms as combat action expends them on the battlefield while spewing huge amounts of carbon and diverse forms of toxic pollution.

The costs of empire to the industrial nations that participate far exceed the material benefits. The political-economic and financial tactics of control via the International Monetary Fund, the World Bank, and the operations of "economic hit men" (Perkins 2004) have achieved far more success in exerting imperial financial control than any military intervention. Yet, the nations of the Global North increasingly find themselves saddled with more debt.

Today, the deployment of new technologies of warfare, private contractors, surrogate soldiers, secret torture sites, clandestine operations, and drone strikes aimed at "high-value" individual and group targets characterize the ever-expanding global "War on Terror." Putin's war against Ukraine offered all sides the opportunity to test and refine these technologies of destruction, and they are. Civilian lives and infrastructure are the greatest losers as Putin tries to militarily deny the existence of a nation.

However, military operations are very expensive. The result of extensive military action is a rapid approach to the empires ruin. With rising nationalism, the US "War on Terror" stimulated the indigenous urge to retain or recapture cultural and territorial sovereignty. Naturally, invasions and occupations have produced resistance, rebellion, fanaticism, and, not so naturally, suicidal terrorism. The unintended consequences of the empires overreach will likely soon trigger its rapid ruin. The domestic culture wars, extreme economic inequalities, political instabilities, and the rise of racist White nationalism and insurrection are all symptoms of the overshoot and approaching ruin of the empire.

Folly of Empire, Failure of Terror

Most Americans remain only semiconscious of the undeclared wars and covert military actions performed in their name around the world. The unspoken mandate of the US military is to enforce the political relations and unbalanced trade of the economic empire. The role of the military in an empire is not as heroic as the actions of its individual members may be while serving in harm's way. One might even view the empire of globalization as the anti-hero of our time.

Intimidation is a core element in the enforcement of empire. When the military enforces the extraction of maximum value from materials and labor by corporations to feed their global economic empire, success often eludes them. The stick of military occupation may augment the smaller carrot of "foreign aid" in achieving political compliance. However, the high stress in conflict zones often leads to an excessive exercise of force through the deployment of "military advisers" and the operations of clandestine Special Forces.

Moreover, the ultimate purpose of interventions and military support for others is not always clear. The moral basis for supporting the Ukrainians in their fight against the blatant invasion and attempted annihilation of people and infrastructure seemed clear and is consistent with the NATO goal of preventing Russian aggression toward Europe. Yet, the new extreme right in Congress took a seeming pro-Russia stance with no apparent purpose other than to oppose any policy of the Democrats.

No one could justify the knee-jerk US support for Israel, despite Netanyahu's seeming fascist ambitions and ruthless slaughter of tens of thousands of Palestinian civilians in retaliation for the Hamas massacre on October 7, 2023. The weak claim that indiscriminate bombing of civilian residential neighborhoods and hospitals is the "right to defend ourselves" does not hold up. The role of the US in the Middle East, which had been predominantly based on the desire to keep open access to its vast oil supplies, increasingly involves ambiguous political goals. Public pronouncements about defending democracy ring hollow.

Military invasion and occupation by no means assure successful control of a subject population. Russia's invasion of Ukraine and the US debacles in Vietnam, Iraq, and Afghanistan have demonstrated the folly of using military force to subjugate a people. It usually guarantees insurgent resistance and often encourages corruption among local political elites in occupied territories, which further stimulates resentment, resistance, and insurgency (Chayes 2015).

Sometimes, political ideology and associated public fear and loathing result in the exploitation of the military for reasons not at all clear in terms of international relations or foreign policy. The fog of war too often begins at the top. The twenty years of US equivocal occupation of Afghanistan ultimately produced little more than widespread food insecurity, deteriorating health conditions, and resurgent Taliban misogyny expressed in the ruthless deprivation and oppression of women and girls.

By design, military institutions are tyrannies. They enforce strict compliance from the top of the hierarchy down, wherever governments deploy them, forcing compliance by both subject governments and people. They do not "bring democracy" to anyone; democracy is an organic social process in which people engage when egalitarian relations are part of their culture.

Such a crass cover story does not change the fact that military personnel on the ground often believe it and do their best to assist the civilian population under nearly impossible conditions. The people involved are merely a means to another end, despite any honorable intentions of occupying personnel. The US, having not fully achieved true democracy itself, is hardly in a position to export its caricature to the world.

The Iraq debacle may be the most obvious example of the self-defeating imperial ambitions of civilian leaders who have little or no military experience and less common sense, amplified by overly intellectualized egos. They left a depleted treasury and caused widespread death and destruction in their wake: think Cheney, Rumsfeld, and, of course, John Bolton. Men with too much power always consider subordinates as inferior or even not-quite persons. Collateral damage in such wars includes the physical and mental wounding of the men and women whose honorable efforts the absurd political ambitions of delusional civilian authorities directly violate.

Never mind for the moment the false premises under which the oil-infused White House administration rammed the invasion and occupation of Iraq through a cowardly Congress unwilling to exercise its constitutional authority to decide whether or not to declare war. When an oil industry committed draft-dodging, Vice President Cheney forced military action destined to fail even in military terms; only disaster for all could follow. Neither Donald Rumsfeld, as the secretary of defense, nor Paul Bremer, as administrator of the Coalition Provisional Authority of Iraq, the ostensible civilian leadership of the Iraq occupation, had even a minimal understanding of how to occupy another country successfully. But they desperately wanted to control that oil.

Rumsfeld's corporate cost-saving mentality cost many military lives by failing to supply what troops needed to protect themselves against improvised explosive devises (IEDs). Bremer destroyed the only viable Iraqi administrative class that could have managed the country under occupation. The only clear consequence of the folly of destroying the social fabric of Iraq was to produce deep resentment and resistance among the people of the occupied country, whose lives were thereby terrorized.

It all led, finally, to insurgency and, ultimately, indigenous terror-ism. It was only because of the US administrative debacle that later, as it all came apart, Al-Qaeda could finally gain a foothold in Iraq— or that ISIS would eventually emerge in part from the radicalized vestiges of the former Iraq military leadership. These absurdities certainly qualify as blowback on a grander scale than even Chalmers Johnson (2000) could have imagined.

The obvious tragedy in the lifespan of most empires is that they inflict suffering and death on the people of the subject nation while draining their own nation's resources and will. That is no less true today. Even more important for the military personnel on the ground, they must perform impossible tasks. Men and women who, after 9/11, believed they had enlisted in a fight to protect their "homeland" from "terrorism" and "bring democracy" to the world found themselves engaged in arbitrary and brutal operations of unclear purpose other than to destroy—and often to kill anything that moves.

Decades earlier, the US attempt to control the destiny of little Vietnam had devolved into a massive, failed exercise in domination through physical intimidation, torture, and indiscriminate killing of civilians. Troops found it difficult to distinguish civilians from insurgent forces resisting the occupation in the south. The US command mistakenly relied on corrupt political puppets, adding to the chaos of failure to capture the "hearts and minds" of the people.

Invasion and attempted occupation bring resistance. In asymmetric warfare, the militarily weaker native peoples have the advantage of being indigenous, which enables them to engage effectively in guerrilla warfare against invading military units unfamiliar with the territory or culture. Hit-and-run assaults terrorize invading troops, after which the attackers disappear into the local terrain and population. So it was in Iraq, Afghanistan, and Vietnam. That, and incredible courage and intellect, along with NATO material support, is why the Ukrainians were able to stave off the Russian invasion so effectively.

Almost everyone who lived through the period of the Vietnam War is aware of the My Lai massacre. Many assume that it was a horrible one-off, anomalous event. In fact, it was iconic of the widespread pattern of indiscriminate killing that resulted from putting ordinary soldiers into the extremely frustrating position of fighting enemy resistance forces they could not distinguish from the civilian population. Nick Turse (2013) documents the facts of which most Americans are still unaware. In that respect, the wars in Iraq and Afghanistan were not very different from the US in the Vietnam War.

Under military occupation, the entire population becomes the presumed enemy. Many soldiers, after regularly seeing their comrades maimed and killed in Iraq by the enemy deployment of improvised explosive devices (IEDs), found themselves committing indiscriminate acts of terror and revenge entirely at odds with their core values. Some violence-prone individuals came to enjoy the power of torturing and killing the powerless and dehumanizing others, as happens in all wars. Most, however, experienced some level of psychological trauma. PTSD haunts them for the rest of their lives. The folly of empire filters down to the individual soldier as personal suffering.

Empires are engines for the accumulation and wielding of both power and wealth. They confiscate and ingest every form of power and economic value while exerting the power of intimidation and violence. The exercise of power may take the form of an administrative state, military occupation, economic extortion, political corruption, or any combination thereof. In any case, the name of their game is to grow more and more powerful to dominate more and more territories in order to maximize control over people and extract resources. Today, the most common form of domination is financial control through lopsided international trade and consequent debt with the assistance of the World Bank and the International Monetary Fund—to the detriment of the people of the target nation.

Historically, various empires large and small have come and gone. They have taken diverse forms. However, no empire can ever escape the limits of the natural systems it attempts to dominate, nor can it violate the laws of physics. Many have tried and failed.

During the fifteenth and sixteenth centuries, Spain and Portugal built empires spanning the globe, beginning with their early voyages of exploration. The British Empire emerged in the seventeenth and eighteenth centuries as the dominant global colonial power, establishing control over the high seas and vast territories, especially in North America and India. With its vast resources and manifest destiny, having transformed itself from a subject colony to an independent expanding nation, the US emerged from the nineteenth century to become the world's leading industrial and military power. For the most part, today's US-led empire of economic globalization does not operate so much by territorial administration or occupation. Rather, it is an empire of economic power, political manipulation, and military threat.

Empires never end well. When faced with their own overextension or with environmental limits, they tend to overshoot; most empires double down on their systemic errors and experience rapid collapse. Some, like the British Empire, slowly acknowledged their loss of control and gave up the political institutions of the empire. This results in a relatively soft landing while clutching the old symbols and economic levers of global authority as they hold on to whatever economic leverage they can. In any case, the linear model of operating an empire ultimately fails in a nonlinear world.

Domination, Resistance, and Control

In recent decades, the "Western powers" have experienced growing resistance to domination by their petroleum-driven industrial empire. Certain "state sponsors of terrorism," such as Iran, sponsor terrorism by "non-state actors." Terrorism, it certainly is. However, the globalists mistakenly attribute the cause of terrorism to religious fanaticism and poverty. Terrorist groups such as Al Qaida and ISIS actually enable, distort, and exploit religious fanaticism in fomenting hatred for the "infidel invaders" in order to help motivate their members to take extreme actions. However, without the aggression of the empire as its target, terrorism would have little to motivate it.

Insurgent groups arise to resist domination by invaders and try to assert cultural and political sovereignty. The Mujahedeen insurgents emerged in Afghanistan to defend against the Soviet Union's invasion and attempt to occupy their country. The Soviets exercised their own imperial terrorism but failed, like so many previous invaders in the history of Afghanistan. Without the Soviet attempt to expand its empire as its target, little motivation for insurgencies would have developed. However, that experience paved to way for the rise of the Taliban in response to the later US invasion and partial occupation.

Terrorist groups like Al Qaida and ISIS took it a step further, fueling their movements with delusions of pseudo-religious grandeur. They encouraged and exploited religious fanaticism as tools to motivate naïve young recruits. Their ruthlessness seems unsurpassed, and they attack "infidels" anywhere rather than merely defending their own "homelands." Such movements feed upon the growing resentment of the empire, which they easily cast as the epitome of evil.

The implicit assumption that insurgent opposition to empire is simply terrorism that arises from the "evil" culture and/or barbarism of a medieval religion has become a convenient illusion perpetuated by the ideology of a petroleum-driven empire. US politicians made no such assumption when the CIA had secretly funded and supported the Mujahedeen in its fight against the Soviet invaders. Once the Soviets were gone, the US, having achieved its immediate goal, abandoned the Afghans to the postwar devastation, causing resentment about exploitation and betrayal. Sadly, Trump's later abandonment of the heroic Kurds, who did the most to defeat ISIS in Syria, repeated the US pattern of betrayal. Then he proclaimed a firm date for US withdrawal, leaving his successor, President Biden, with little choice but to follow through or risk the massive complications of a new escalation by occupation forces.

Originally, the US invasion and initial occupation of Afghanistan were ostensibly meant to hunt down Osama bin Laden in the aftermath of the 9/11 attacks. Yet, the oil-crazed White House quickly turned to attack Iraq on trumped-up charges of having weapons of mass destruction and imaginary terrorist affiliations. A Congress too weak and gullible after 9/11 to make its own decisions passively accepted these false narratives.

Meanwhile, the Mujahedeen morphed into the fundamentalist Taliban and, with other insurgent groups, rose up to resist the ambiguous US occupation weakened by the redeployment of the bulk of US forces to Iraq. If it seems hard to find rationality in all this, remember we are dealing with the illusions of empire on the part of Texas oilmen rather than any clear, fact-based, goal-seeking strategies rationally executed with concrete purpose. The goals of US policy toward Afghanistan remained a moving target over two decades of mixed anti-terrorist and anti-insurgency operations punctuated by furtive gestures toward "nation-building" amid rampant corruption (Chayes 2015).

If we were to believe the illusions of empire—and so many do—the Taliban fighters of Afghanistan simply rose up from nowhere to attack its civil government and its US and NATO allies, supposedly there to "assist" in the establishment of democracy. In fact, as mentioned above, the Taliban arose from the remnants of the Mujahedeen, the indigenous insurgents that the US had sponsored as surrogate forces used in the 1980s to drive the Soviets out after they invaded Afghanistan.

The CIA had covertly supplied the Mujahedeen fighters with billions of dollars in arms and cash in its attempt to create for the Soviets their own Vietnam-like failure. Like so many invaders throughout Afghanistan's history, the Soviets withdrew in 1989, unable to overcome the native insurgents. The Afghans likely would have thrown out the Soviets even without US support, as it had so many other invaders throughout its history. Once the Soviets were gone, the US lost interest in Afghanistan until Bin Laden hid out there after 9/11.

Like the Soviets and the British before them, the Americans also were unable to subdue the Taliban or stabilize Afghanistan after two decades of partial occupation. US ambiguous intentions there did not help either the Afghans or the troops on the ground. President Biden removed the last US fighting forces in mid-2021, leaving a political-military vacuum as the Taliban gained ground in several provinces. Soon, its factions quickly controlled the whole country.

Through the exploitation of distorted fundamentalist notions of true religion, some resistance movements (insurgencies) morphed into cults of pure violence with the central intent to terrorize "infidels" everywhere. We saw their works on the nightly news, just as they intended. They deployed delusional forms of self-glorified martyrdom, exploiting their impressionable and heavily propagandized disaffected youth in the role of suicide bombers.

The obvious examples are Al-Qaeda and later the so-called Islamic State of Iraq and Syria (ISIS), also known as The Islamic State of Iraq and the Levant (ISIL), or as Daesh, its Arabic language acronym for the name of the Islamic State: al-Dawla al-Islamiya fi al-Iraq wa al-Sham, a pejorative term for ISIS in Arabic (Reston 2016). With illusions of establishing a new Caliphate on the Arabian Peninsula, ISIL emerged from the remnants of resistance to the US invasion and its unresolved occupation of Iraq. Its battlefield effectiveness results from leadership that had once been part of the Iraq military, which the US had foolishly disbanded years before. Irony abounds.

The growing cult of violence against the forces of the empire grew to formerly unimaginable levels of terrorist acts, expressing and amplifying a generalized culture of hatred. Yet, what is not terrorizing about an AGM-114 Hellfire missile reigning down death and dismemberment on a village from unseen drones in the sky? What are cluster bombs falling down upon villages in Yemen, indiscriminately maiming and killing civilians, if not terrorism?

What, but terror, resulted from Special Forces teams invading the homes of Iraqi families late at night, often killing residents based on flimsy "intelligence" by paid informants that a "suspected terrorist" might be present? To retain our semiconscious pretensions to both empire-denied and "defending freedom and democracy," US authorities applied double standards to their terror and ours. The ultimate result of such illusions of cause and effect is to accelerate the global catastrophic convergence of poverty, political instability, starvation, migration, and violence with the escalating climate and ecological destabilization caused by the global-industrial project of the empire itself.

On 9/11/2001, a few Saudi nationals committed acts of spectacular death and destruction in New York City and Washington, DC. They used our high technologies fueled by Middle East oil—large US jet airliners and skyscrapers—to destroy both and themselves. They made the threat of terrorism very real for Americans by bringing it directly to the symbols of American financial and military world dominance, the World Trade Center and the Pentagon.

In the shock of those events, the irony of their turning our technologies against us escaped most Americans. Whatever incompetence of the US intelligence services or presidential failures to respond to important intelligence briefings available prior to the attacks only amplifies that irony. Hubris may have played a role in the failure to recognize the signs of the impending attack. Two decades later, the indifference of IDF leadership in Israel to intelligence reports of the impending Hamas massacre on October 7 is a striking parallel of hierarchical hubris among colonial and imperial powers.

An entire literature of conspiracy theories has emerged in attempts to explain how a rag-tag band of mostly Saudis with minimal technological understanding could combine exploitation of our high technology with fanaticism to bring down the financial center of the American empire. The official 9/11 Commission remains widely perceived to have covered up government indifference, if not complicity or even conspiracies, as an explanation for the political failures to act on available intelligence.

Richard Clarke (2004) was the national coordinator for Security, Infrastructure Protection, and Counterterrorism for the United States under several presidents. He offers a disturbing account of the Bush administration's consistent indifference to the daily security briefings he gave George W. Bush. The briefings, based on multiple intelligence sources, warned of probable impending attacks by Al Qaeda not long before the 9/11 attacks. Clarke's prescient warnings that an invasion of Iraq would be a major national mistake are also striking.

Illusions of technological and cultural superiority might also have shattered, yet, they persist. For some, only if it had been an inside job could American superiority have become so vulnerable and violently breached. Personally, I would not have imagined such a gambit beyond Dick Cheney's deranged vision of autocratic power seeking a pretext for war. However, mastering the operational complexity of pulling off such a conspiracy would seem beyond Cheney's capacity, if not his desire, making it highly unlikely.

Recognition of the wisdom of the iconic cartoon character Pogo, "We have met the enemy, and he is us," remains as elusive as national self-reflection is rare. It is much easier to retain simple illusions of linear cause and effect than to acknowledge political and economic culpability in creating the complex problems from which we suffer. Social control for both national security and societal wellbeing continues ungrounded in realistic assessments of the conditions around us and our role in creating them. The more recent ambivalent support among Americans for defending Ukraine against Putin's barbaric and criminal attack on its people revealed a similar lack of political clarity.

Collapse

The United States of America has been at war for most of the two-and-a-half centuries of its existence, including no less than 106 conflicts (Wikipedia, n.d.) One report said that the US has been at war 225 out of 243 years since 1776 (Shah 2020). Yet, with the exception of the recent terrorist attacks and the many wars waged against the native populations throughout the era of Westward expansion, none of those wars reached American soil. It is hard to argue that the United States is not militaristic. Certainly, some of its many wars were justifiable, even principled. Nevertheless, the US conducted its wars and military actions on the soil of other nations or on the high seas around the world, far from the "homeland." The US initiated many on shaky grounds, characterizing its actions elsewhere as somehow defensive.

The post-9/11 trope, the "homeland," elevated to the status of cabinet-level Department of Homeland Security, is a big clue. It suggests an extended vision of the global reach and international intentions of the US political elite. The framing of the euphemism "War on Terror" suggests offensive action against a largely unbounded and undefined enemy. Words have social power; they frame our emotions about the experiences or ideas to which they refer. Combined, the terms "War on Terror" and the "homeland" lend emotional support for an aggressive militaristic foreign policy without limits. Together, they suggest an image of unlimited incursions into every corner of the planet in a quest for total control everywhere to eliminate terrorism and protect the "homeland."

The defining features of the "War on Terror" are its global reach and its unrestrained but ambiguous targeting. The US political class has a penchant for fighting rhetorical and political wars on everything from poverty to drugs, almost universally without winning. However, as traditionally defined, terrorism is a tactic of violent resistance, usually to the rule of dictators—also by dictators—or to invasion and occupation, or exercised by an invader deploying violence against civilians who resist domination. Its goal is to weaken either political authority or resistance. It often works both ways.

Terrorism is a practice, not a nation, territory, or group. Reference to the "homeland" implicitly defines the entire rest of the world as the battleground on which America fights its "War on Terror." The idea entails a silent recognition that the terror may reach our own territory if we do not attack it elsewhere. Well, it is too late for that now.

To what does "homeland" refer to in the political-military context of America looking out at the rest of the world? By not specifying a distinct national enemy against which to wage war, which is, of course, hard to do when many terrorists are non-state actors, the terrorism meme can refer to anyone, anywhere, other than the citizens and territory of the US. The term offers a propagandistic legitimacy for carte blanche permission to assert military power and commit covert (or even overt) military violence anywhere, anytime.

A presidential kill list, despite having no legal precedent or authority, naturally followed. John Brennan, President Obama's Homeland Security advisor, created it, and Obama used it extensively. Its euphemistic formalization, the expanded "disposition matrix," an extensive global database of suspects marked for assassination, including US citizens, achieved official status as a permanent foreign policy tool (Miller 2012). Not surprisingly, Trump escalated its aggressive use around the world by not requiring his direct approval of an action. Upon his inauguration, President Biden put into place the requirement that nobody would be put onto the list without his approval. Yet, the list persisted.

It is important to note here the strange relationship between the "War on Terror" and the "War on Poverty" in US foreign policy. Foreign aid has a long, ambiguous role in US relations with non-industrial nations. The parallels and interconnections are striking. Foreign aid has been a frequently used tool to exert political influence over other governments. The evidence shows that most individuals who committed the terrorist acts on 9/11 and others were educated and not poor. Yet, the US political elites have defined poverty as an important root cause of terrorism. They define the people of poor countries as characteristically violent, which is a political case of denial of responsibility and an inaccurate projection of cause onto others.

Official US policy frequently links an international "War on Poverty" to the "War on Terrorism." Politicians use that link as an excuse to redirect foreign aid to certain nations that, by the World Bank's own measure, have the world's worst governments. This failed ideology of fixing failed states rests on the illusion that poverty causes terrorism, and we must prop up ruthless regimes in order to develop their economies and thereby avoid terrorism. The policy fails on both counts. Private foundations, such as that of Bill and Melinda Gates and World Vision have contributed hundreds of millions to this failed effort. Dictators often divert the aid to their own coffers while inflicting state terrorism on their own people (Easterly 2016).

When I first heard the term the "homeland" used by the neoconservative interventionist Cheney/Bush White House crowd, I felt a creepy sense of deja vu. They wanted to invade Iraq, claiming it was a response to the destructive attack on symbols of American power by a motley band of extremist Saudis. Yet, they knew that Al-Qaeda's leader who organized the attack, Osama Bin Laden, a wealthy Saudi, was hiding in Afghanistan. Iraq had not been involved. Saddam Hussain hated Al-Qaeda.

At the time, I did not remember that the Nazi propaganda machine had used the term "homeland" to suggest the superiority of a "German race" and to differentiate it from the rest of the world. However, the term did bring to mind images of Nazi Stormtroopers invading Poland and France in World War II and conquering other nations for the glory of the "homeland."

How, exactly, did this "homeland" trope suddenly arise in the context of the misdirected US response to the 9/11 terrorist acts? The Nazis had referred to the "homeland" extensively in their attempts at racist self-glorification in publicly framing their desire to rule the world. The fact that "defending the homeland" abruptly appeared in the American political lexicon was no accident. Defending the US from terrorist attacks was one thing. Invoking a Nazi trope to defend international aggression that was clearly not defensive is quite another thing.

It struck me as emblematic and very peculiar that the mass media adopted the term "homeland" so quickly and uncritically. It had popped up suddenly in the post-9/11 Cheney/Bush rhetoric, and the cable pundits and network news anchors adopted it immediately without question. By then, of course, the corporate media had become little more than a conduit for the drumbeats of war. But war against whom?

The desire to strike back at an undifferentiated foreign "them" was a natural reaction for most Americans after 9/11. I overheard terms like "towel head" and "sand nigger" on the street, racist references to ambiguous evil "others" out there. Ambiguously targeted racism resonated with the nation's fear and anger at a much-generalized image of the otherwise diverse peoples of the Islamic world. Unfortunately, too many Americans indiscriminately applied racist tropes to anyone, such as Sikhs, who might "look like them."

Because Saddam Hussein was such a ruthless dictator, Iraq became an easy target simply by using a slick propaganda campaign to paint him as a supporter of international terrorism. The neocons skillfully steered their desire for revenge to their imperious desire to conquer oil-rich Iraq. Defense of the "homeland" became a powerful meme justifying any military attack of their choosing anywhere in the world.

As it turns out, reflection on some of the more clandestine elements of US foreign policy, especially from the latter half of the twentieth century and beyond, is entirely consistent with the more recent developments post-9/11. Anyone who has read Naomi Klein's book *The Shock Doctrine* (2007) should understand how the globalized political economy of perpetual growth of capital guides so much of US foreign policy. Disaster capitalism, as Klein describes it, has created and exploited indebtedness and subservience in so-called underdeveloped or developing nations around the world. Of course, the two terms "terrorism" and "homeland" function to disguise the fact that such nations have been both the victims of and material sources for the enrichment of the corporate elites of the nations of the Global North ever since the very first conquistadors.

US political administrations, both Republican and Democrat, have consistently deployed indirect terror through surrogate dictators in many nations. US actions, usually covert, have placed in office and supported ruthless dictators such as Pinochet in Chile and Suharto in Indonesia in exchange for political-economic subordination and financial plunder. In Honduras, the US State Department led by that interventionist corporate Democrat, Hillary Clinton, quickly legitimized the 2009 military coup that overthrew the democratically elected President Manuel Zalaya. Death squads and criminal gangs drove terrorized peasants to flee, many to the United States, where growing numbers of xenophobes denigrate the refugees of US foreign policy as "criminals, rapists, and terrorists." US support for military coups in Latin America is nothing new.

Indirect terror through surrogate dictators has pervaded US policy in Latin America and elsewhere for a long time. Clandestine threats of terror through economic hitmen, such as described by John Perkins (2004), have required political and economic subservience to US corporate economic interests through enforced indebtedness supported by political, trade, and military policies. Perkins wrote from his own direct experience in the strange corporate-government hybrid role as just such an economic hitman.

As Perkins reports, economic hitmen attempt to persuade political leaders to take on huge debt to build mega-infrastructure projects of little, if any, benefit to their people—using American contractors and equipment. If their persuasion fails, shadowy higher-ups call in the jackals to resolve the problem. The uncooperative leader turns up dead in a mysterious plane crash or another disguised form of assassination. In one form or another, force sustains the global empire of corporate economic growth, further driving the planet into the catastrophic convergence of crises: poverty, starvation, migration, war, and both ecological and climate destabilization.

Terror can come in many forms, from the brutal slaughter of women and children by conquistadors of old to assassinations by modern Special Forces operators. Terror ranges from the blatant beheadings of innocent journalists by Middle Eastern jihadists to the reigning down of US Hellfire missiles on villages or wedding parties executed by drone pilots half a world away.

"They" (those we define as terrorists) define imperialist invaders of their land and associated contract killers (for example, Blackwater) as terrorists who use the technologically advanced forces of the most powerful industrial nations in the world. Our "insurgents" are their "freedom fighters." The threat of ISIS and other fanatical jihadists remains very real and very dangerous, but where did it start? We could look all the way back to Vasco da Gama's first voyage in 1497–1499 to the East and his hatred of Muslims and desire to conquer and destroy them all. Barbaric bloodshed followed and has continued in that vein for centuries. However, the efforts of the industrial nations to secure control over the world's oil supplies and other resources needed for industrial growth since the twentieth century are today's most important precipitants of contemporary state terror. Despite the obvious and urgent necessity of weaning ourselves off fossil fuels, big oil companies initiate new petroleum exploration and extraction projects. They do so to feed the failing project of endless economic growth with the help of continuing government subsidies.

Two forms of terror dominate the world today. The first is the terror of the corporate state forcing its expansion at the expense of ordinary people and environments everywhere. It is the terror of plundering capital, which extracts resources, disrupts ecosystems, spills wastes, and oppresses indigenous peoples around the world. The second is the terror of insurgent groups resisting the empire of corporate growth with increasingly fanatic brutality, in part by fomenting and exploiting distortions of medieval religious beliefs as their excuse for indulging in unbridled violence.

However, terrorism, once started, takes on a life of its own. Its self-amplifying bloodlust is addictive and generates its own ideological justifications. No one can justify any form of terror without wild stretches of the imagination and an outright denial of facts. The modern cycle of international violence escalates without reference to human values, as the lust for power and the dehumanization of the foreign "other" drives it.

Most Americans are ignorant of the fact that extended droughts, violent gangs imported from the US, and government-sanctioned death squads drove the growing rates of immigration from Honduras, Guatemala, and El Salvador. These nations were also plagued by drought and food insecurity. Political opportunists define innocent immigrants fleeing terror as "potential terrorists." The White nationalist hatred and vilification of immigrants encouraged by such demagoguery continues to grow in Europe, the US, and the Middle East, so long as the root causes remain unchallenged.

Ultimately, a new form of terror arises when amplified by the global forces behind these two kinds of terror. It is the terror that arises from having to face unlivable conditions caused by ecological destabilization, economic instability, climate chaos, population displacements, political upheaval, armed violence, and approaching societal collapse. Human destruction of planetary stability is causing this new form of terror.

The terrifying deterioration of the conditions of life today has only just begun. Most people perceived the record-breaking destructive heat waves of the summer of 2023 as anomalous, freak events. However, others recognized that they would soon be seen as relatively cool compared with future global heating. Only by massive intervention to restructure the relations among societies and by transforming the role of humanity on the planet can we begin to realize a potentially less terrorizing future. That will require restructuring our minds.

A strange but beautiful symmetry emerges when we consider the real-world practical solutions to the increasing devastation of the planet by the ecological terrorism of the corporate state. Add to that the political-militarist terrorism of the jihadists, the emerging terror of growing climate chaos, and the imminent societal collapse inherent in the continued heating of the planet. Remarkably, the solution to all of them is the same, though extremely complex and difficult to achieve.

When we consider the planetary consequences of continuing on the multiple intersecting (and terrifying) paths that make sustaining human life on Earth extremely problematic, we cannot deny the necessity of a radical turn toward a new strategy for human survival. We must recognize that the business-as-usual operations of the global corporate-industrial empire, rising jihadist terrorism, endless wars of choice, and emerging resource wars amid mass migrations all increase the risk of the extinction of the human species.

Empire and terror are two primary forces directly threatening human survival, not just by the bloodshed involved but also by their impact on both the planet itself and our ability to respond to the current New Great Transformation in any creative way. War may be the biggest polluter of all. Putin's invasion and war of aggression on Ukraine, for example, destroyed not only a great deal of that nation's civilians, its infrastructure, including homes, schools, hospitals, and power plants. It also contributed vast quantities of carbon pollution to the Earth System, further accelerating the race toward both the Earth System and societal collapses. Without some form of very positive and comprehensive human intervention, the pursuit of empire will continue to be a major contributor to climate and ecological chaos, which can lead to societal collapse.

We must redirect many contemporary forces of ecological destruction if we are to survive. This includes applying new forms of creative destruction to societal institutions and practices, including terror and war and excessive industrial production, shipping, consumption, and waste, if we intend to overcome and replace the multiple forms of planetary destruction we have caused. Both empire and terror directly obstruct humanity's ability to transform our relations with the planet and each other, and they reduce our chances to survive through this century and beyond.

CHAPTER NINE

Social Ecology
After Hierarchy

COORDINATION AND CONTROL in societies is a complex matter. This book is titled *Holding It Together* because coordination and control have become increasingly tenuous and have even lost a clear sense of purpose as we approach the end of the Industrial Era. That is partly because few recognize the emerging processes of massive social change and ecological and climate disruptions that confront us. Actually, material disruptions—wildfires, super storms, droughts, floods, crop failures, and increasing societal disturbances—have already affected the everyday lives of many people around the world.

We can no longer take for granted our ability to hold it together in the world that industrial modernism has so radically changed. The forces of resistance to changing the status quo would have us go along and get along in the conventional ways. The environmental modernists pretend that all the deep systemic disruptions are mere "problems" that we can "solve" within the existing systems and institutions. Societies can no longer hold it together on that basis, which is why we must radically restructure them just to survive the first half of the twenty-first century.

Clearly, the facts contradict the ideological optimism of the defenders of the failing structures of authority. The existing hierarchies, veiled in pseudo-democratic forms, cannot even recognize the nature of the great transformations of ecosystems and climate that the endless-growth economy has caused but cannot resolve. That is why humanity is so far behind in responding to the greatest existential threat ever to confront us.

The dominators of this world will not give up controlling the system that has made them super-rich. It is up to the rest of us to initiate the New Great Transformation of society that is required in order to mitigate the extreme damage and destabilization of the Earth System that the system they dominate continues to cause. In this final chapter, we examine the potential and the necessity of the New Great Transformation of the global-industrial-consumer economy into a viable ecological civilization.

Massive societal transformation must occur if we are to salvage a livable Earth System in which humans and other species can survive. We, the people of planet Earth, cannot wait for our "leaders" to lead; we must force them to follow us. They are not as wise as Mahatma Gandhi, who said, "There go my people. I must follow them, for I am their leader."

From Detached Hierarchy to Engaged Democracy

The fundamental underpinnings of the Western and now global industrial-consumer culture, despite their temporary material power expressed in the Industrial Age, are in deep conflict with the basic operating principles of the natural world in which we live. Here is the short version.

Plato and his dualistic philosophy began the journey, from ancient Greece to medieval society in Europe and the early Christian church, then incorporated into the Renaissance and the new science and technology that ultimately became the basis for the Industrial Revolution. Plato believed that the mind—by which he meant cognitive thought—was divine and enduring and that the body was vile, impermanent, and secondary. That fundamental assumption has continued through history, and it drives the cultural and psychological illusion of separation of humans from Nature today. Yet, we live in our bodies, and our bodies live in the world.

The history of human consciousness, as it evolved in the West, in contrast with other cultures, makes its reductive dualistic thinking not only fascinating but also deserving of serious consideration. This is especially true because of the great impacts it has had on the world and, ultimately, because of its destructiveness. Jeremy Lent (2017, 2021) wrote two powerful books that document in great detail and insight humanity's search for meaning through the ages. Lent explains how the dominance of dualistic and reductionist thinking has alienated modern humans from our own Nature.

The predatory, extractive approach to our relations with Nature and other humans has caused immeasurable harm to the Earth System and to human relations. The associated hierarchy and technology have not only produced riches for some and oppression for many more, but they have also caused severe damage to our own living habitat, the Earth System. Anyone who wants to understand the deep roots of the alienation of modern humans from our own existence, and its consequences, should read Jeremy Lent's books.

Political, economic, and societal hierarchy is the essential structure of the culture of predation and domination most modern people have come to see as normal. While some reciprocity exists in hierarchies, the structure of power in modern societies has little or no grounding in broader mutual support in society. Its central principle is domination—one-way, top-down control. Modern industrial-consumer culture attempts to dominate Nature, the very source and substance of our existence on planet Earth.

That is why the dominator culture has trampled on ecosystems and indigenous cultures across the globe, with such cavalier indifference and cruelty, ever since the earliest European exploration, conquests, and colonization of peoples all around the world. Because of the power of its technology and the ruthlessness of its dominator mentality, the world became a rigid hierarchy of peoples, with the nations of the Global North on top.

That arrangement, which still dominates the world today even as its form has changed, has increasingly damaged the health and well-being of not only humans but also the broad spectrum of life on our planet. The trajectory of hierarchical domination is both self-perpetuating and unsustainable, simply because it is steadily and at a rapidly increasing rate, destroying the very foundation of life within the Earth System. It cannot and will not stop itself.

Nevertheless, there is a difference. Today, the global transformation is not just about how society is organized. It involves first the ongoing severe disruption of the natural balance sustained by the Earth System itself during the Holocene Epoch now apparently ending. The destabilization of the Earth System accelerates—caused by the compounding of the processes and products of the Industrial Age—disrupting the planetary life-systems of which we humans are a part. Our survival obviously depends on the health and stability of the biosphere, climate stability, and the continuity and stability of ecosystems around the world. The continuation of the global-corporate-political economy of perpetual economic growth threatens them all.

The only way we can stop, or at least constrain, the destruction, is by changing our ways in the most fundamental sense. That is, we must abandon the global political-economic system that is destroying Earth-System stability and replace it with one that can live in harmony with the life-supporting elements of Nature that we are rapidly destroying. At the same time, we must repair the damage we have already done, in part, by repairing society itself.

That means abandoning the increasingly extreme hierarchical political and economic structures that have caused the mess we are in; those structures are incapable of changing themselves. We must replace them with societal formations, both political and economic, that distribute power equitably among people and communities who work in their local and regional contexts to live in harmony with Nature and with each other. However, we must urgently repair, restore, and regenerate the complex living systems—our ecosystems—upon which we ultimately depend for survival.

Most of the efforts so far to initiate actions to stabilize the Earth System have focused on getting national and global institutions—consumers, corporations, and the governments that support them—to stop relying on fossil fuel energy to operate their enterprises. Climate activists, in particular, have struggled courageously in their efforts to persuade the world's most powerful bankers, fossil-fuel executives, corporate CEOs, and "opinion leaders" to move away from fossil fuels as their primary source of energy for their operations.

But that cannot happen in any significant degree without major societal reorganization. The utter failure of decades of the UN's Conferences of the Parties (COPS) is obvious. The last one before this sentence was written—COP28 (held in 30 Nov – 13 December 2023) and headed by an Au Dhabi oil company executive, Dr. Sultan Ahmed Al Jaber, appointed president of COP28 while head of Abu Dhabi National Oil Company (ADNOC), was the epitome of international hypocrisy.

Even with some progress in installing solar and wind generation of electricity, global carbon emissions have only grown. This is partly because clean energy generation has *added to,* instead of replacing, dirty power plants, even as some coal-fired power plants have shut down because they are no longer economically competitive. All the while, total energy consumption has grown. So we have even more total carbon emissions, atmospheric and oceanic heating, and ecological destruction.

That is because the existing economic system of endless growth remains untouchable. This is an example of Jevon's paradox, where technology and/or policy add supply (in this case, renewable energy) at a reduced cost, which induces increased demand, resulting in more total consumption.

While we have made some small, mostly symbolic progress in recent years, corporations and even governments have grown quite adept at greenwashing—feigning cooperation while refusing to change their basic business model. However, the trajectory of the accelerating growth of carbon emissions from the global political-economic system far exceeds the pace of real change in pollution patterns. Something much more comprehensive and complete must happen. So far, it has not because the belief in economic growth remains sacred and seems politically untouchable.

Consider the annual meetings of the global elites of wealth and power at Davos, Switzerland. The World Economic Forum membership consists of mostly old White, wealthy leaders of corporations, governments, and financial elites from the high-income nations of the Global North—the billionaire class (Goodman, 2023). While paying lip service to climate action, these self-appointed "masters of the universe" continue consolidating their wealth and power, assuring the continued growth of their empires.

Peter S. Goodman, a global economics correspondent for *The New York Times,* spent a good deal of time among the billionaire class in order to write his book *Davos Man: How the Billionaires Devoured the World* (2022). He described how their deployment of capital to capture distressed assets such as housing after the 2008–2009 financial collapse and then driving up rents financially devoured the global economy. Naomi Klein (2007) rightly calls that strategy "disaster capitalism". Such economic power drives government power not just via lobbying but also by financial influence and legal authority. But there is another level of extremely centralized social control by the billionaire class.

Claire Provost and Matt Kennard have taken a different approach to mapping the rise of the global corporate empire. Allocation of resources and the governing of societies are largely a matter of private corporate decisions over the commercial and financial direction of both commerce and government. They describe the *Silent Coup* (2023) executed by the billionaire class, which is now what can only be called an oligarchy hiding behind a hollow shell of democracy.

As investigative journalists for the Centre for Investigative Journalism in London, Provost and Kennard traveled to thirty countries where they talked to people on the ground and discovered how the corporate empire penetrates and controls ordinary people's lives. The corporate empire exerts social control globally through intricate systems of corporate welfare (never called that). They have forced legislation to establish privately controlled international tribunals that can overrule the laws of any sovereign nation in order to protect profits and even potential profits of extractive industries such as coal, gold, or rare earth minerals. This strategy has insulated them from democratic decision-making and public oversight. Always, the people suffer the loss of resources, economic value, control over their lives, and often their health.

To work effectively for relative climate stability—compared to our increasingly unstable climate today—and human and ecological health and survival, we must overcome the global corporate empire. By we, I mean we, the people, and not some cinematic White hero on a white horse wearing a white hat. It is far more complicated than seeking heroes to save us. As Bill Moyers put it a while back, only organized people can overcome organized money.

The unfortunate fact is that we live in a world in which most of us are not only heavily indebted to but are, in one way or another, entangled with the very system we must fundamentally change in order to survive as a species. Those with institutional power in the hierarchy are more obligated and committed to the system than to the rest of us. However, to reestablish our relationship with the Earth System—our habitat—society must comprehensively reorganize itself. That means people reorganizing their institutions and the way we live. No existing "leadership" is either capable or willing to take on such a seemingly impossible task.

The people must democratically and creatively organize new societal formations to reflect a high awareness of the Earth System itself and each of its component living systems—we must act collectively on that awareness. It is all about us, not some "great-man" theory.

However, we still live under the same hierarchy that is the cause of our global predicament. Any real solutions can come only when societies overcome hierarchy to engage the world and each other democratically and equitably. Only when communities make decisions based on assuring the wellbeing of their people instead of further entrenching the power of the billionaire class to accumulate more capital, to the detriment of everyone else, will the outlook become more hopeful.

In order to achieve democratic control of national and international political economics, serious cultural and social change is necessary. So far, change seems to be going in the wrong direction. Certainly, there is much discontent, but too much of it is manipulated by the political-economic elites to blame vulnerable scapegoats—immigrants, Black folks, etc. Meanwhile, the political economy keeps chugging along as if it need not change its practices. On top of that, billionaires fund major propaganda efforts, feeding recently rising ethno-nationalism in most wealthy developed nations, which produces more exclusion and oppression of diverse ethnic minorities while dodging the fundamental structural failure of the global system.

Perhaps the most iconic example is the impact of the history of racism in America, where White nationalism and political hierarchy amplify growing economic inequities along ethnic and racial lines.

That is why we must look closely at the cultural and structural necessities for societal justice and empathy as we experience even more environmental degradation and instabilities moving further into the Anthropocene. The new societal transformation we need will not work unless driven by democratic participation and compassion for all sentient beings—I mean that in a very concrete way. That worldview is nowhere in the playbook of the dominant culture of the political economy, which is precisely why it has to go.

Parameters of Societal Survival

The average member of the billionaire class holds a worldview very different from the rest of us. However, it would be a mistake to believe that all those riches make the super-rich much happier than everyone else. Lots of research makes clear that money makes people happier only to the extent that it relieves the pain of poverty and deprivation of the basics of life and helps overcome health and other problems of life. Beyond that, happiness involves much more important things, such as how we relate to other people and to the rest of the world. So, what is it about the growing billionaire class? In a culture of domination, the core values of power and greed tend to not only prevail among the most powerful but also infect many others, some of whom even emulate those who oppress them.

The rich *are* different than the rest of us; they have much more money and vastly more political power. Questions of status, security, and interpersonal relations are rooted in the same factors, although they play out differently at the top of the societal hierarchy, where financial power is the dominant theme of life. The rich and powerful are rich and powerful because they obsess over wealth and power, with the possible exception of some who fell into wealth by inheritance and are not part of the game—they are merely high-end consumers.

The highest value in the worldview of the billionaire class consists of the ruthless acquisition of more and more money and power.

That is why it is so difficult to get the attention of the most powerful men in the world— most are old White men, after all—when it comes to making the decisions that they now must make (or someone must make for them by removing their power) if humanity is to survive in the very near future. They are unwilling to give up control of the complex techno-economic processes that use so much fossil-fuel energy, despite the fact that those processes continue to destabilize the Earth System. Members of the billionaire class live in a wealth-and-power status bubble the rest of us could never afford (or want) to occupy or maintain. Now, none of us can afford for them to have so much wealth and power since that is the core source of the destabilization of the Earth System that has sustained us until now.

A combination of democratic politics, scientific knowledge, and indigenous wisdom is our last best hope for forging a viable path to human survival amid the increasing global chaos now accelerating into the unknowns of the Anthropocene. There are signs of hope in many places, yet too many trends are moving in exactly the wrong direction while the good trends are moving much too slowly.

For example, in 2022, governments around the world gave a total of about seven *trillion* dollars in subsidies to fossil fuel companies, even as those same corporations made record profits. Much of it has gone to investments in exploration and development just when we need to keep more oil and gas in the ground. Most of the billionaire class still acts as if nothing needs to change, except maybe for ordinary people to recycle more plastic packaging if they want to buy any lettuce at all.

Consider how climate scientist and communicator Katharine Hayhoe sees it: "What if we got rid of these subsidies? The IMF estimates that would prevent 1.6 million premature deaths annually, raise government revenues by $4.4 trillion, and put the world on track to reaching global emissions targets. All of this would result from the cleaner air, less lung and heart disease, and fiscal freedom that would follow cutting and properly taxing carbon dioxide emissions." (Hayhoe 2023) Dr. Hayhoe is no wild-eyed radical; she is a seriously religious, humanitarian scientist.

It will not be easy to do what we must do no matter how well anyone behaves. Yet, the most powerful people in the world continue behaving very badly—while making gestures of caring via the deceptions of philanthrocapitalism (Shiva 2015). The Davos Men (Goodman 2023) make proclamations of "responsible climate policy" while they focus their real attention on accumulating more wealth in their hedge funds and diverse asset portfolios, regardless of the social and ecological costs.

If the rich and powerful were on board with reality outside their bubble of imaginary power over Nature and society, it would be so much less difficult to develop and execute plans to restore so much of the Earth System. We could transform societies to minimize the damage from past and present carbon emissions and ecological destruction, thereby maximizing our chances to survive and flourish in the coming decades and centuries. However, those at the very top of the global hierarchy are not on board with fundamental change. They have fought for decades to grow and maintain their hegemony over the world. They have mostly succeeded; that must stop.

A New Great Transformation of the global political economy, which is necessary to stabilize climate and ecosystems, would require a major redistribution of power from the current extreme hierarchy to a much flatter network of ecological communities. That lies at the core of what Vandana Shiva (2020) has referred to as the necessary "resurgence of the real." Shiva refers to the reality of Nature, where we all reside, including the billionaire class.

Not only does excessive hierarchy—with far too much power consolidated at the top of the "political-economic pyramid"—stifle attempts at ecological recovery by deploying powerful propaganda designed to confuse the public. Unfortunately, social instabilities add to the cultural confusion, amplifying the fear, anger, and eventually, the hate encouraged by political demagogues. Growing social conflict also results from geographic and internet mobility in relation to increasing threats to people in diverse geographic locations. Many fear for their lives and migrate to escape climate chaos, starvation, political oppression, and war in order to enhance their chances of survival.

Democracy is a delicate process that requires constant maintenance. As long as hierarchy is the most prominent structural form in a society, democracy will remain very difficult to establish or maintain. After all, democracy requires knowledgeable citizens engaged in public life and political actors who commit to the rule of law and the will of the people.

A "representative democracy" is not a democracy unless it represents the people and is not subservient to ruling elites in a rigid hierarchy. On that score, we have a long way to go. The vast majority of legislation in most industrialized nations, especially in the US, serves the interests of the most powerful political-economic elites, not the people.

Adding fuel to the fire of failure, politicians of hate increasingly turn to voter suppression and intimidation. They know their policies are opposed by the majority, who recognize that they are not served well (or at all) by the political and financial elites, even when confused by the deliberate distractions of scapegoating more vulnerable groups. Societal survival requires several things. One is harmony. Identification with other beings and our Earth System is essential. Willingness to work toward mutual support and economic organization based on maintaining the wellbeing of the people requires compassion.

Historical and anthropological evidence demonstrates that cooperative forms of non-hierarchical, or minimally hierarchical, networks have predominantly organized themselves and survived for millennia.

Today's swollen population of gluttonous, detached, and alienated consumers is entirely out of sync with the character of needed changes and with the core processes of the Earth System. That is why societal survival can only result from a New Great Transformation of the global political economy and family, community, and nation—and their culture. The global political economy, controlled by the super-rich, many of whom are members of the World Economic Forum, currently dominates societies today. That is precisely why progress on climate and ecological action has gotten nowhere.

These mega-elites attend their annual meetings at Davos to plan how they can retain control of the globalized economy that enriches them to the detriment of all other living creatures and systems. At the same time, they are quite adept at pontificating about the importance of responding to global crises, and they proclaim strategies and funding for actions that never quite come to pass.

Ecological civilization is the only human social formation that can enable to survive much further into the Anthropocene. An ecological civilization is the exact opposite of the high-energy consumption, high-extraction production, high-consumption, high-waste, and highly destructive global political economy we experience today as a "normal" modern life. The industrial-consumer culture resists even contemplating such a radical change, as is implied by the idea of an ecological civilization. Meanwhile, the Davos Men continue to control the political and economic parameters within which the industrial-consumer society operates.

Culture resides in the minds of the members of society. The industrial-consumer culture frames for us a life outside of Nature, as if we were somehow independent of the very world in which we live.

That is no more realistic than the mainstream economist's belief that we can somehow overcome the finite limits of planet Earth by inventing new technology and creating new materials to replace the ones we are rapidly depleting by the magic of "progress through chemistry."

The very first step toward securing human survival at this late date is to curtail severely the burning of fossil fuels, which supply energy for so many operations of the industrial-consumer economy and of our everyday lives. That is almost common knowledge now among the relatively informed. Yet, I can find almost no public discussion of the implications of radically reducing fossil-fuel energy use for the organization of societies that depend on high energy use for their excessive industry and consumption. Suggestions as to how to reorganize the economy to operate on a low energy-use basis never accompany pleadings to reduce carbon emissions, and fossil fuel usage are never accompanied.

The evidence, in fact, is quite strong that we must reduce carbon emissions to near zero (not the loophole-filled "net zero") to stave off the worst overheating of the planet. While 1.5° or 2.0° Celsius above pre-industrial average global temperatures would seem small out of context, those are just averages. In fact, we will probably pass the 1.5° Celsius level before you read this. Knowledge is one thing; action is quite another. Because of decades of inaction, it now appears overshooting these "targets" is fast approaching realization. Already, as we approach 1.5° C., the acceleration of frequency and intensity of extreme weather events is obvious.

Much more extreme temperatures are already happening in some regions, but it is not simply a matter of changing temperatures. The increased energy in the atmosphere and oceans causes major, complex systemic changes in climate patterns. More powerful hurricanes and more erratic tornados, accelerated breakup and melting of Arctic and Antarctic ice, causing feedback loops and sea rise, and other catastrophic climate complications continue to proliferate. What the climate skeptics do not understand is that we are looking at very complex systems that have remained in relatively stable balance for the eleven thousand years or so of the Holocene Epoch. They provided the conditions conducive to the development of modern humans and our complex civilizations. These conditions are not "givens"; they are already changing rapidly.

From the beginning, the strategies of our industrial modernists did not consider the changes in the Earth System caused by their predation on Nature in building our complex industrial societies. Nor, in most cases, do they yet. Early on, it was mostly a matter of ignorance. Now, it is a matter of refusal to face facts. However, we have reached a tipping point, a dead end beyond which the ability of the living Earth System to tolerate our behavior quickly evaporates. Without severe intervention on our part to change radically our relations to our only home, Planet Earth, the collapse of industrial civilization, seems inevitable. It will be either collapse or fundamental transformation.

So, at its core, it is a very simple matter, yet, it requires incredibly complex collective actions on the part of the people of the industrial-consumer societies, that is, the nations of the Global North. On the one hand, cutting carbon emissions seems quite simple: just stop all burning of fossil fuels in all sectors of the global-industrial-consumer economy. On the other hand, imagine how complicated transitioning away from fossil fuels will be. Fossil fuels energize virtually every modern economic, political, and social institution and activity.

Everyone in the Global North, and a much smaller portion of the peoples living in the Global South, operate their economies on the energy derived from burning fossil fuels. Everyone uses some kind of fuel; some use far more fuel far more intensely than others do. Think of Jeff Bezos and Elon Musk; they qualify as the symbolic tip of the overheated iceberg of energy waste.

Remember, oil, gas, and coal are the original solar energy storage depositories. They were part of the processes that provided enough oxygen in the Earth's atmosphere to support the lives of an expanded fleet of species, culminating in us mammals, and finally enabled human flourishing. We have been replacing our life-giving oxygen with carbon for several hundred years, and we now feel the early consequence: the catastrophic changes in climate and ecosystems that are now underway. Our survival requires that we reverse that process.

However, the culture of industrial production and consumption is deeply entrenched in our heads and in our "built environment." Virtually all modern infrastructure is fossil-fuel dependent. Even the so-called "service economy" relies on fossil fuels in many ways, from the buildings occupied by, maintained by, and repaired by various workers who also use high-energy transportation systems to go to and from work.

It does not take much thought to realize that every product and devise, and everyone in industrial civilization, depends on fossil fuels, directly or indirectly, to support their operations and livelihood. Elimination of our dominant energy source will require massive reorganization of society, and that means drastically changing how we live.

The irony in this great human predicament is that whether or not humans will step up and proactively take the very difficult steps necessary to transform our global high-energy political economy into low-energy-consuming, ecologically sustainable communities, the emerging New Great Transformation of human society will happen. It will occur in either a good way or a very bad way, with very different outcomes.

If we fail to act decisively, as we have so far, the destabilization of the living Earth System that sustains us will destroy industrial civilization. If we muster the political will and cultural creativity to mobilize enough people to force the transformation of society in the process of eliminating fossil fuels, we may have a chance to build a place of our own on Earth in the form of local/regional ecological communities. Even if we do so in the very most effective ways, the process will be very messy and painful. It is up to us if it is more or less so.

The force of Nature's response to industrial civilization will transform us because of our failure (so far) to respond to the parameters for human survival, which are inherent in Nature and are not subject to our will. If we humans do not act together to move forward with the necessary changes in how we live in relation to the world around us and in relation to each other, Nature will take its course anyway—and the consequences for humans will not be pretty.

Holding it Together While Going from Here to There

So, what will it take for us to hold it together while trying to forge a new way of living in harmony with Nature and each other? Well, a lot of things. Many of those who resist the idea that industrial civilization must be transformed or massively downsized to achieve a civilization using vastly less power to manufacture goods and supply services misconstrue the result. They imagine that as only going back to living in caves and carrying clubs. Well, they forget two things—which, by the way, they always tout as the reason we don't have to change our ways—the innovative skills and the technical knowledge we already have and the creativity that we can apply to living in very different ways now.

There are, of course, limits to innovation in the development of new technologies and products. Those are material limits, the kind of thing Donella Meadows and her colleagues predicted so accurately in 1972. However, I am talking about cultural innovations applied to restructuring our lifeways to replace fossil fuel-driven, industrial-scale production with low-energy technologies and organizational forms.

That, of course, will also involve economic de-growth, equitable redistribution of assets, and elimination of many of the superfluous products and processes enabled by fossil-fueled industry but less than peripheral to achieving a good life. That will also require a great deal of cultural creativity. The business of humanity must reinvent itself.

If we had no cultural creativity, then transitioning out of industrial-consumer culture to create an ecological civilization would be extremely tough. Of course, a major change in the very structure of society will be very difficult in any case. But the social innovations we now need involve applying our creativity to our social relations with each other, in order to transform our relations with Nature for our survival. Once the public realizes that the motivation to create new forms of social relations to fit into a low-energy consuming ecological civilization will be very high. That is where the oft-touted "entrepreneurial spirit" can flourish.

At the same time, people don't just go around willy-nilly changing their culture. One of the major advantages that human groups built in the process of social evolution was the ability to recognize what worked and what did not work and to remember to keep doing what worked. The big difference is that in the hunter-gatherer days, our ancestors lived in very stable small-scale environments in which they could survive by learning and doing what their elders had done for generations. They almost never had to deal with social change within the span of several lifetimes, no less one generation. We, on the other hand, have caused and adjusted to a great deal of social change in recent centuries. Now, we face the necessity to create the greatest social change ever, even more quickly, to form an ecological civilization from the dregs of a dying global-industrial-consumer political economy.

By comparison with past evolutionary changes, the Industrial Era has played out over a very short period—just a few centuries. In that time, and to a lesser extent before and during the agricultural revolution, many things changed in major ways. Humans have had to adapt to rapid change during and since the Industrial Revolution (Polanyi 1971[1944]). However, it was not so much that the natural environment changed. The application of new technologies to greater control over Nature and over other peoples by emergent political hierarchies changed the world. Now, we must develop and deploy new technologies, some very powerful old ones, and new forms of social organization to create ecological communities that operate in harmony with their habitats.

We have faced major changes throughout the Industrial Era. The enforcement of hierarchies and the subjugation of "others" caused great conflicts within and among societies unlike anything our hunter-gatherer ancestors experienced. That continued throughout the Industrial Era, but now, something fundamentally different has occurred. Nature has caught up with our predatory extractive practices as we approach the end of the Industrial Era, and we must face the consequences of our not-always-creative destruction of the living Earth systems upon which we depend.

At the same time, the human instinct to participate in community remains an essential core component in our long-evolved character. Consider the global response to major natural disasters. As I first wrote a rough draft for this chapter, rescue agencies had identified over twenty thousand victims and counting of a devastating series of earthquakes in Turkey and northwest Syria. The international response provided strong evidence of the survival of the human tendency to feel compassion and to express the value of community, even across national borders and political rivalries during a humanitarian crisis.

Even the government of Ukraine, under siege by Putin's bloody autocratic attempt to obliterate that nation, sent professional search-and-rescue teams to help in the rescue effort. Human solidarity and the rendering of mutual aid can override even the most rancorous political rivalries. Not much later, we saw evidence of residents of Lahaina, Maui, running into the burning city to rescue neighbors they may not have even personally known. The alienation that characterized the industrial-consumer "individualism" has not extinguished the human penchant for empathy and compassion. As the list of catastrophic events and unnatural disasters continues to grow, the core human value of mutual aid will continue. It will be an important element in the struggle to transform societies to achieve economic systems based on wellbeing and harmony with Nature.

Similar evidence at every scale of human organization indicates that the fundamental human characteristic of community feeling can overcome and transcend hierarchies. It expresses direct human compassion by actions supporting those in need without any reward other than the satisfaction and meaning derived from helping others. We will never find meaning and purpose in life in the next consumer product (or the next billion in profits), although the propagandists of commerce would have you believe so. And we will not survive without reinvigorating the natural human instinct to organize ourselves for mutual support.

Holding it together while undergoing what I expect will be the greatest societal transformation in human history is possible because of three things.

First, humans have retained that evolutionarily successful trait of cooperation grounded in compassion and human solidarity. When people recognize the suffering that results from a growing crisis, most will feel compassion for others and will cooperate to relieve that suffering. The power elites try to keep us separate as "individuals." They cannot.

Second, when humans recognize that the contemporary ways of living in relatively static hierarchies no longer work, they will come to realize that we must be creative and engage in societal innovations to meet the existential crises that confront us.

Third, when the emergency is full-blown, most people, not all, will rise to the occasion and take the necessary actions, even heroic, unprecedented acts, to respond to the emergency, no matter how dire.

Now, all this may sound a bit idealistic, but it is more the practical necessity of overcoming oppressive conditions when you consider the extreme difficulty in making those three things happen under the present global-political context. Many sociological studies of collective behavior have demonstrated that under certain conditions, humans suspend the normal rules of society to respond to emergencies.

Most such studies have focused on short-term disasters. However, we are in a "long emergency," the term James Kunstler (2005) used to refer to the converging catastrophes of climate and the other emerging crises of the twenty-first century. Nevertheless, I believe the same principles apply, as evidenced, for example, by the continuing committed and coordinated mutual aid in the struggle of the Ukrainian people to overcome the continued brutality of Putin's war crimes. The key here is applying creativity to self-organizing for survival. That principle can be equally applied to our global long emergency, which will take much longer and even more resourcefulness to overcome than a finite, even horrendous, regional war.

While a vast quantity of evidence demonstrates that humanity is already in a climate/ecological emergency, a variety of factors mitigate against many people recognizing that fact. The oil industry and its allies' decades of spending billions of dollars on propaganda every year, and the general tendency to resist change have amplified the complicated convergence of all sorts of political and economic short-term interests in retaining the status quo. Yet, more and more people are recognizing that the emperor wears no clothes.

Many of the reasons for the collective resistance to facing ecological and even economic facts of system failures, such as those discussed in this book, also cause long delays and institutional resistance to change. We delay recognition, creativity, and action to the point where it may very soon be too late for the right actions to have enough effect to stave off the unmitigated collapse of civilization itself. Yet, self-transformation of civilization is the unavoidable necessity of the twenty-first century. To paraphrase Chris Hedges regarding fascism, we fight climate chaos not because we are sure to win but because it is climate chaos.

I am sure that once sufficiently severe catastrophic events become so obvious in their direct consequences that the collapse of civilization is clearly imminent, people will rise up everywhere and try to do the right thing. But, by then, I fear, it may actually be too late to stop the collapse of all the institutions, supply lines, and production of necessities that ordinarily sustain human life. The COVID-19 pandemic was a minor blip on the economic screen by comparison. At that point, community self-sufficiency and mutual aid will be harder to initiate and sustain. As the saying goes, the best time to plant a tree was twenty years ago; the second-best time is today.

Even by taking the most direct and rational action now to stop burning fossil fuels and reorganize societies around low-energy technology and an ecological economy, we will not be able to avoid great displacements and suffering. However, it is more likely that things will not go that well since we have yet to mobilize to make such global changes. If it comes to that, vast numbers of people will die of a variety of causes tied to the collapse of industrial civilization caused by continued Earth-System destabilization combined with the break-down of institutions. In that case, human populations will shrink by attrition through starvation, destitution, disease, and armed conflict.

But that does not have to happen. Admittedly, we are already at a point where accelerating food insecurity, starvation, political and armed conflict, and massive climate-forced displacements are already happening in some places. Yet, there is still time to avoid much chaos and suffering if people begin to rally and organize to take direct action to stop fossil fuel burning immediately. That has to be priority one; we simply cannot afford to put more fossil-fuel exhaust into the world.

Of course, many industrial processes that damage ecosystems require fossil fuel energy, so their destructiveness would also decline significantly. At the same time, however, we must reorganize society to live with the aid of low-energy technologies, many of which already exist, while simultaneously working hard to repair, restore, and regenerate the ecosystems on which we all depend for survival.

I cannot imagine a taller order. Yet, when confronted with necessity, humans have demonstrated that we can and will cooperate to achieve mutual aid when needed and to change the way we live in order to survive—it is our evolutionary heritage. We can, if we will, hold it together in order to get there from here. That is what hopeful realism is all about.

Acknowledgments

So many forces and people influence the development of a book that it is difficult to sort out the impact or benefit of each. Therefore, I shall list here only those kind and thoughtful individuals who have directly contributed to my work on the book itself. I am grateful for the many comments and suggestions made by those patient and kind enough to have read early drafts. Far more helpful than I ever imagined were the 'beta-readers' who read, commented, critiqued, and made suggestions on the earlier much larger manuscript that on advise of my editor I agreed was too dense, too long with too many related topics. So I extracted a set of relatively coherent material to rewrite into *Holding It Together*. I thank these important contributors in the next paragraph.

Earl Kessler brought the experience of a lifetime in community development and disaster relief work with various agencies and NGOs around the world, to bear on his reading of the early work, enriching the final product to the extent that I was able to incorporate them. Jan Deligans' comments and suggestions were invaluable. Derek Jones brought his down to earth sociological perspective and insights from both street-level community life and deep academic experience to bear on important suggestions for improving this book. Darby Southgate is a consummate sociologist whose comments and suggestions reflected her sociological rigor. Dexter "Ed" Bryan, a career-long colleague and friend, offered helpful suggestions, as well as the admonition that I really ought to get back to surfing in Southern California.

Larry L. Rasmussen, Reinhold Niebuhr Professor of Social Ethics emeritus at the Union Theological Seminary, provided some of the most important suggestions, which helped me make improvements to the book, and kindly wrote the Forward. Jack Heinsius, long time business economist, and fine human being offered incredibly incisive comments and detailed suggestions especially for reorganizing chapters. Richard

Nasef, entrepreneur, inventor, family therapist, and poet, made a number of important comments on an early draft. Esha Chioccio, a talented photographer and tireless climate activist in Santa Fe, offered valuable suggestions on readability in reaching an audience. Larry Kilham, a successful technologist, entrepreneur and prolific writer and poet, offered insights from an economic perspective.

Sarah Lovett, best-selling fiction author, editor, and book coach offered very useful suggestions on more effectively telling the story of the greatest challenge to ever face humanity. Keith and Mish Schneider, former floor traders and life-long market analysts offered insights on economic and market factors of great benefit to me. Our extensive conversations on today's global situation have provided important context for my writing. Wayne Underwood, writer, philosopher of the Southwest, and former television and film technical engineer offered important insights into the human condition. Dave Thomas' personal experience, from warfare on the other side of the world to religion and everyday life in village communities in Mexico, align well with his insightful humor, all of which benefitted the revisions of my first efforts that led to this book.

Freelance copyeditor Jen Z. Marshall's important copyediting prevented me from violating a variety of rules delineated in the *Chicago Manual of Style* and from committing unlimited typos. Nick Zelinger's talent as a book designer is obvious; just look at the cover and the inside formatting of this book, or refer to his design awards. While I thank each of these colleagues, friends, and diverse professionals for the very real benefits of their suggestions and comments on my work, they are not responsible for any errors, omissions, or misinterpretations I may have made in the book. They are mine alone.

My work has also benefitted from connections with diverse experts and professionals in fields like climate science, evolutionary biology, ocean conservation, neuroscience, social ecology, and social psychology, as well as environmental activism and more. Some of them are among the prepublication reviewers I am also grateful to for agreeing to review the book. They include Erin Remblance, Antoinette Vermilye, Earl Kessler, Derrick Jones, Jack Heinsius, Mish Schneider, Mark Asquino, Mark Van Patten, Pamela Pence, and James Richards, who kindly took the time to review and comment on the Advance Reader Copy of *Holding It Together* on fairly short notice.

Finally, I am most deeply grateful for the highly talented work of my editor, partner, friend, and spouse, Cynde Van Patten Christie, who is far more responsible than I am for any clarity, effective organization, or unpretentious expression the reader might find in my book. Her professionalism is unparalleled, which reflects well in the best-seller status of many of the authors whose work she has edited. Nevertheless, I take full responsibility for the substance of the argument and whatever disappointment the reader may find in this book.

References

Bardi, Ugo. 2017. *The Seneca Effect: Why Growth is Slow but Collapse is Rapid: A Report to the Club of Rome*. Cham, Switzerland: Springer International Publishing.

Brown, Clair. 2017. *Buddhist Economics: An Enlightened Approach to the Dismal Science*. New York: Bloomsbury Press.

Buettner, Dan. 2012. *The Blue Zones, Second Edition: 9 Lessons for Living Longer from the People Who've Lived the Longest*. Washington, DC: National Geographic.

Butler, Smedley D. 1935. *War is a Racket: The Antiwar Classic by America's Most Decorated Soldier*. Port Townsend, WA: Feral House, 2014.

Carlevale, Edmund. 2023. Post on LinkedIn on Carbon Capture and Sequestration. (In reference to Dr. Charles Harvey's Talk at the Cornell Atkinson Center for Sustainability.) https://www.linkedin.com/feed/update/urn:li:activity:7121553207897530368/

Catton, William R. 1980. *Overshoot: The Ecological Basis of Revolutionary Change*. Urbana: University of Illinois Press.

Catton, William R. 2008. "A Retrospective View of My Development as an Environmental Sociologist." *Organization and Environment*. (21: 4, 471-477).

Centola, Damon. 2021. *Change: How to Make Big Things Happen*. New York: Little, Brown Spark.

Chayes, Sarah. 2015. *Thieves of State: Why Corruption Threatens Global Security*. New York: W. Norton.

Childs, John Brown. 2003. *Transcommunality: From the Politics of Conversion to the Ethics of Respect.* Philadelphia: Temple University Press.

Christie, Robert MacNeil. 2014. "Becoming Indigenous: Settling a Population Adrift in an Unstable World." Resilience: Building a World of Resilient Communities. (December 24). https://www.resilience.org/stories/2014-12-24/becoming-indigenous-settling-a-population-adrift-in-an-unstable-world/

Clarke, Richard A. 2004. *Against All Enemies: Inside America's War on Terror.* New York: Free Press.

Coates, Ta-Nehisi. 2015. *Between the World and Me.* New York: Spiegel and Grau.

Coram, Robert. 2002. *Boyd: The Fighter Pilot Who Changed the Art of War.* New York: Back Bay Books.

Crowley, Roger. 2015. *Conquerors: How Portugal Forged the First Global Empire.* New York: Random House.

Diamond, Jared. 2005. *Collapse: How Societies Choose to Fail or Succeed.* New York: Penguin Books.

Duhigg, Charles. 2012. T*he Power of Habit: Why We Do What We Do in Life and Business.* New York: Random House.

DeFries, Ruth. 2021. *What Would Nature Do? A Guide for Our Uncertain Tunes.* New York: Columbia University Press.

Ferguson, Niall. 2018. *The Square and the Tower: Networks, Hierarchies, and the Struggle for Global Power.* London: Penguin Books, UK.

Forrester, Jay W. 1968. *Principles of Systems.* Cambridge: Wright-Allen Press.

Gates, Bill. 2019. *Breakthrough Energy Coalition.*
http://www.b-t.energy/coalition/

Gates, Bill. 2021. *How to Avoid a Climate Disaster: The Solutions
We Have and the Breakthroughs We Need.* New York:
Alfred A. Knopf.

Gessen, Masha. 2020. *Surviving Autocracy.* New York: Random House.

Gibben, Edward. [1776] 1995. *The Decline and Fall of the Roman
Empire.* New York: Modern Library.

Gilens, Martin, and Benjamin Page. 2014. "Testing Theories of
American Politics: Elites, Interest Groups, and Average Citizens,"
Perspectives on Politics 12:3 (September):564-581.

Goodman, Peter S. 2022. *Davos Man: How the Billionaires Devoured
the World.* New York: Custom House (Harper Collins).

Gowdy, John M. 2021. *Ultrasocial: The Evolution of Human Nature
and the Quest for a Sustainable Future.* Cambridge, UK:
Cambridge University Press.

Greer, John Michael. 2011. *The Wealth of Nature: Economics as if
Survival Mattered.* Gabriola Island, BC: New Society Publishers.

Harvey, Charles. 2023. "Why Carbon Capture and Sequestration is
Better for Producing Oil than Fighting Climate Change."
Lecture at the Cornell University Atkinson Center for
Sustainability. (September 7)
https://www.youtube.com/watch?v=KObewsFHA1U

Hayhoe, Katharine. 2023. "Not So Good News." (September 1)
Katharine Hayhoe Newsletter on LinkedIn.
https://www.linkedin.com/pulse/green-energy-saves-day-katharine-hayhoe/?mid
Token=AQH18GOPVBxgJw&midSig=2HAGxLVRuzIWU1&trk=eml-
email_series_follow_newsletter_01-newsletter_hero_banner-0-

open_on_linkedin_cta&trkEmail=eml-email_series_follow_newsletter_01-news
letter_hero_banner-0-open_on_linkedin_cta-null-2mfd8h~lm0ul6bi~25-null-
null&eid=2mfd8h-lm0ul6bi-25&otpToken=MTMwMDE2ZTYxMjJ
lYzFjZGIzMjQwNGVjNGYxZGUyYmM4N2NlZDU0NTkxYTg4ODYxNzRjM
TA1Njc0YTVlNThmNWYzZGZiMDllN2JjY2VmZDY2NjljOTNkYWYxOTA2
MzdhMjM0MzViNGNjMGU4ZTZjMTk2NjUsMSwx

Heinberg, Robert. 2011. *The End of Growth: Adapting to Our New Economic Reality*. Gabriola Island, BC: New Society Publishers.

Hickel, Jason. 2021. *Less is More: How De-Growth Will Save the World*. London: Penguin Random House UK.

Johnson, Chalmers. 2000. *Blowback: The Costs and Consequences of American Empire*. New York: Metropolitan Books.

Johnson, Chalmers. 2004. *The Sorrows of Empire: Militarism, Secrecy, and the End of the Republic*. New York: Henry Holt.

Kelly, Marjorie. 2023. *Wealth Supremacy: How the Extractive Economy and the Biased Rules of Capitalism Drive Today's Crises*. Oakland: Berrett-Koehler Publishers.

Klein, Naomi. 2007. *The Shock Doctrine: The Rise of Disaster Capitalism*. New York: Metropolitan Books.

Kunstler, James Howard. 2005. *The Long Emergency: Surviving the End of Oil, Climate Change, and Other Converging Catastrophes of the Twenty-First Century*. New York: Grove Press.

Lent, Jeremy. 2017. *The Patterning Instinct: A Cultural History of Humanity's Search for Meaning*. Lanham, MD: Prometheus Books.

Lent, Jeremy. 2021. *The Web of Meaning: Integrating Science and Traditional Wisdom to Find Our Place in the Universe*. Gabriola Island, BC: New Society Publishers.

Lenton, Tim. 2016. *Earth System Science: A Very Short Introduction.* Oxford, UK: Oxford University Press.

Marshall, John "The Hunters." 1957. Documentary film directed in collaboration with Robert Gardner, shot on a Smithsonian-Harvard Peabody sponsored expedition in 1952-53. See: https://en.wikipedia.org/wiki/The Hunters (1957_film) Several other documentaries have shown the lifeways of these people and the effects of the incursions of moderns into their territories. See: https://en.wikipedia.org/wiki/San_people.

McNeill, J. R., and Peter Engelk. 2016. T*he Great Acceleration: An Environmental History of the Anthropocene since 1945.* Cambridge, MA: Harvard University Press.

Meadows, Donella, Dennis Meadows, Jorgen Randers, and William Behrens III. 1972. *Limits to Growth.* New York: Universe Books.

Meadows, Donella. 2008. T*hinking in Systems: A Primer.* Edited by Diana Wright. White River Junction, VT: Chelsea Green.

Mills, C. Wright. 1959. *The Sociological Imagination.* New York: Grove Press.

Nørretranders, Tor. 1999. T*he User Illusion: Cutting Consciousness Down to Size.* New York: Penguin Books.

Orlov, Dmitry. 2016. *Shrinking the Technosphere: Getting a Grip on Technologies that Limit our Autonomy, Self-sufficiency and Freedom.* Gabriola Island, BC: New Society Publishers.

Partanen, Anu. 2016. *The Nordic Theory of Everything: In Search of a Better Life.* New York: Harper Collins.

Perkins, John. 2004. *Confessions of an Economic Hit Man.* San Francisco: Berrett-Koehler Publishers.

Polanyi, Karl. 1971 [1944]. *The Great Transformation: The Political and Economic Origins of Our Time*. Boston: Beacon Press.

Provost, Claire, and Matt Kennard. 2023. *Silent Coup: How Corporations Overthrew Democracy*. London: Bloomsbury Academic.

Raworth, Kate. 2018. *Doughnut Economics: Seven Ways to Think Like a 21st-Century Economist*. White River Junction, Vt: Chelsea Green.

Reich, Robert. 2015. *Saving Capitalism: For the Many, Not the Few*. New York: Knopf.

Reston, Laura. 2016. "World Leaders Have Taken to Calling ISIS 'Daesh,' A Word the Islamic State Hates," New Republic. https://newrepublic.com/minutes/123909/world-leaders-have-taken-to-calling-isis-daesh-a-word-the-islamic-state-hates

Richardson, Katherine, Will Steffen, et al. 2023. "Earth Beyond Six of Nine Planetary Boundaries," Science Advances 9:17 (13 September) https://www.science.org/doi/10.1126/sciadv.adh2458

Rockström J, Steffen W, Noone K, et al. 2009. Planetary Boundaries: Exploring the Safe Operating Space for Humanity. *Ecology and Society*. 14: art32.

Schumacher, E.F. 1973. *Small is Beautiful: Economics as if People Mattered*. Vancouver, B.C.: Hartley and Marks, 1999.

Schumpeter, Joseph. [1942] 1975. Capitalism, Socialism, and Democracy xs. New York: Harper.

Selye, Hans. [1956] 1978. *The Stress of Life*. New York: MacMillan.

Shah, Sabir. 2020. "The US Has Been at war 225 out of 243 years since 1776." *The New International*. (January 9)

https://www.thenews.com.pk/print/595752-the-us-has-been-at-war-225-out-of-243-years-since-1776

Shiva, Vandana. 2015. *Soil Not Oil: Environmental Justice in an Age of Climate Crisis.* Berkeley: North Atlantic Books.

Smith, Philip B. and Manfred Max-Neef. 2011. *Economics Unmasked: From Power and Greed to Compassion and the Common Good.* Totnes, Devon, UK: Green Books.

Snyder, Timothy. 2017. *On Tyranny: Twenty Lessons from the Twentieth Century.* New York: Tim Duggan Books.

Stanley, Jason. 2020. *How Fascism Works: The Politics of Us and Them.* New York: Random House.

Steffen, Will, Johan Rockström, et al. 2018. "Trajectories of the Earth System in the Anthropocene." *Proceedings of the National Academy of Science (PNAS).* 115:33 (August 14) 8252-8559.

Tainter, Joseph A. 1988. *The Collapse of Complex Societies.* Cambridge, UK: Cambridge University Press.

Thompson, William Irwin. 1971. *At the Edge of History: Speculations on the Transformation of Culture.* New York: Harper & Row.

Thunberg, Greta. 2021. Keynote speech before the Youth4Climate Conference. Milan. https://www.youtube.com/watch?v=ZwD1kG4PI0w

Weber, Max. [1905] 1963. *The Protestant Ethic and the Spirit of Capitalism.* Translated by Talcott Parsons. New York: Charles Scribner's Sons.

Weber, Max. 1947. *The Theory of Social and Economic Organization.* Translated by A.M. Henderson and Talcott Parsons. New York: Simon & Schuster.

Wikipedia. (N.D.) *List of wars involving the United States.* https://en.wikipedia.org/wiki/List_of_wars_involving_the_United_States

Wolin, Sheldon. 2008. *Democracy Incorporated: Managed Democracy and the Specter of Inverted Totalitarianism.* Princeton: Princeton University Press.

Wrangham, Richard. 2009. *Catching Fire: How Cooking Made Us Human.* New York: Basic Books.

About the Author

Robert MacNeil Christie grew up in Southern California during America's economic "Great Acceleration" in the 1950s. Working summers in construction from age fourteen, he soon realized that the smart young Black men who were his mentors on the job had far less opportunity than any 'white boy.' On graduating from high school he pondered the abundant opportunities in the classified section of the *Los Angeles Times*. After a brief stint in college, then in the military, he returned to the university to pursue his education through the PhD, searching for answers to the mysteries and tragedy he saw in so much human behavior.

Dr. Christie taught social psychology, social organizations, research methods, and social change for thirty-five years, while conducting community research with his students in the South Central Los Angeles area. He continues to consult with non-profit community groups on matters of strategy, social change, and environmental issues. Dr. Christie is Professor Emeritus of Sociology and Founding Director of the Urban Community Research Center, California State University, Dominguez Hills. On retiring from teaching, he intensified his investigation of the intersections of social, economic, and political causes of the converging crises of climate, ecological, economic, and political systems, which threaten human survival. That research resulted in his book, *Holding It Together: Social Control in an Age of Great Transformation.*

Always engaged with Nature and technology, after having earned all those merit badges to become an Eagle Scout in his youth, Dr. Christie later became an instrument rated pilot, an Aikido Shodan, a skilled wood worker, and conservationist. He also designed and built two energy efficient homes. He blogs at https://thehopefulrealist.com/ on topics related to converging critical issues involved at the end of the Industrial Age, including climate, ecosystems, political economy, culture, and everyday life. He works on strategy with non-governmental organizations whose missions involve mitigating the greatest challenges ever to confront humanity—the transformation of both Earth System and societies as we move deeper into the Anthropocene.

Index

www.ingramcontent.com/pod-product-compliance
Lightning Source LLC
Chambersburg PA
CBHW052110030426
42335CB00025B/2923